NEURAL NETWORK MODELING

Statistical Mechanics and Cybernetic Perspectives

Perambur S. Neelakanta
Dolores F. De Groff

Department of Electrical Engineering
Florida Atlantic University
Boca Raton, Florida

CRC Press
Boca Raton   Ann Arbor   London   Tokyo

Library of Congress Cataloging-in-Publication Data

Neelakanta, Perambur S.
  Neural network modeling: statistical mechanics and cybernetic perspectives / by Perambur S. Neelakanta and Dolores F. De Groff.
      p.   cm.
  Includes bibliographical references and index.
  ISBN 0-8493-2448-2
  1. Neural networks (Computer science)   I. De Groff, Dolores F.   II. Title.

QA76.87.N388 1994
006.3--dc20   94-14052
                                                                                                CIP

No claim to original U.S. Government works
International Standard Book Number 0-8493-2448-2
Library of Congress Card Number 94-14052
Printed in the United States of America   2 3 4 5 6 7 8 9 0
Printed on acid-free paper

Dedicated to our parents

Contents

Chapter 6: Stochastical Dynamics of the Neural Complex

Chapter 8: Informatic Aspects of Neurocybernetics

Appendix A: Magnetism and the Ising Spin-Glass Model

Preface

Neural network refers to a multifaceted representation of neural activity constituted by the essence of neurobiology, the framework of cognitive science, the art of computation, the physics of statistical mechanics, and the concepts of cybernetics. Inputs from these diverse disciplines have widened the scope of neural network modeling with the emergence of artificial neural networks and their engineering applications to pattern recognition and adaptive systems which mimic the biological neural complex in being "trained to learn from examples".

Neurobiology which enclaves the global aspects of anatomy, physiology, and biochemistry of the neural complex both at microscopic (cellular) levels and at macroscopic structures of brain and nervous system constitutes the primary base upon which the theory and modeling of neural networks have been developed traditionally. The imminence of such neural models refers to the issues related to understanding the brain functions and the inherent (as well as intriguing) self-adaptive control abilities of the nervous system as dictated by the neurons.

The cognitive and learning features of neural function attracted psychologists to probe into the intricacies of the neural system in conjunction with similar efforts of neurobiologists. In this framework, philosophical viewpoints on neural networks have also been posed concurrently to query whether machines could be designed to perform cognitive functions akin to living systems.

Computer science *vis-a-vis* neural modeling stemmed from the underlying computational and memory capabilities of interconnected neural units and is concerned with the development of so-called *artificial neural networks* which mimic the functional characteristics of the web of real neurons and offer computational models inspired by an analogy with the neural network of the human brain.

Since the neuronal structure has been identified as a system of interconnected units with a collective behavior, physicists could extend the concepts of statistical mechanics to the neural complex with the related spin-glass theory which describes the interactions and collective attributes of magnetic spins at the atomic and/or molecular level.

Yet another phenomenological consideration of the complex neural network permits modeling in the framework of cybernetics* which is

---

* The concepts of cybernetics adopted in this book refer to the global self-organizing aspects of neural networks which experience optimal reaction to an external stimulus and are not just restricted to or exclusively address the so-

essentially "a science of optimal control over complex processes and systems".

Thus, modeling neural networks has different perspectives. It has different images as we view them through the vagaries of natural and physical sciences. Such latitudes of visualizing the neural complex and the associated functions have facilitated in the past the emergence of distinct models in each of the aforesaid disciplines. All these models are, however, based on the following common characteristics of real neurons and their artificial counterparts:

- A neural network model represents an analogy of the human brain and the associated neural complex — that is, the neural network is essentially a neuromorphic configuration.
- The performance of a neural network model is equitable to real neurons in terms of being a densely interconnected system of simple processing units (cells).
- The basic paradigm of the neural computing model corresponds to a distributed massive parallelism.
- Such a model bears associative memory capabilities and relies on learning through adaptation of connection strengths between the processing units.
- Neural network models have the memory distributed totally over the network (*via* connection strengths) facilitating massively parallel executions. As a result of this massive distribution of computational capabilities, the so-called *von Neumann bottleneck* is circumvented.
- Neural network *vis-a-vis* real neural complex refers to a *connectionist model* — that is, the performance of the network through connections is more significant than the computational dynamics of individual units (processors) themselves.

The present and the past decades have seen a wealth of published literature on neural networks and their modelings. Of these, the books in general emphasize the biological views and cognitive features of neural complex and engineering aspects of developing computational systems and intelligent processing techniques on the basis of depicting the nonlinear, adaptive, and parallel processing considerations identical to real neuron activities supplemented by the associated microelectronics and information sciences.

---

called *cybernetic networks* with maximally asymmetric feed-forward characteristics as conceived by Müller and Reinhardt [1].

The physical considerations in modeling the collective activities of the neural complex *via* statistical mechanics have appeared largely as sections of books or as collections of conference papers. Notwithstanding the fact that such physical models fortify the biological, cognitive, and information-science perspectives on the neural complex and augment a better understanding of underlying principles, dedicated books covering the salient aspects of bridging the concepts of physics (or statistical mechanics) and neural activity are rather sparse.

Another lacuna in the neural network literature is the nonexistence of pertinent studies relating the neural activities and the principles of cybernetics, though it has been well recognized that cybernetics is a "science which fights randomness, emphasizing the idea of control counteracting disorganization and destruction caused by diverse random factors". The central theme of cybernetics thus being the process automation of self-control in complex automata (in the modern sense), aptly applies to the neuronal activities as well. (In a restricted sense, a term *cybernetic network* had been proposed by Müller and Reinhardt [1] to represent just the feed-forward networks with anisotropic [or maximally unidirectional], asymmetric synaptic connections. However, it is stressed here that such cybernetic networks are only subsets of the global interconnected units which are more generally governed by the self-organizing optimal control or reaction to an external stimulus.)

This book attempts to fill the niche in the literature by portraying the concepts of statistical mechanics and cybernetics as bases for neural network modeling cohesively. It is intended to bring together the scientists who boned up on mathematical neurobiology and engineers who design the intelligent automata on the basis of collection, conversion, transmission, storage, and retrieval of information embodied by the concepts of neural networks to understand the physics of collective behavior pertinent to neural elements and the self-control aspects of neurocybernetics.

Further, understanding the complex activities of communication and control pertinent to neural networks as conceived in this book penetrates into the concept of "organizing an object (the lowering of its entropy) ... by applying the methods of cybernetics ...". This represents a newer approach of viewing through the classical looking mirror of the neural complex and seeing the image of future information processing and complex man-made automata with clarity sans haziness.

As mentioned earlier, the excellent bibliography that prevails in the archival literature on neuronal activity and the neural network emphasizes mostly the biological aspects, cognitive perspectives, network considerations and computational abilities of the neural system. In contrast, this book is intended to outline the statistical mechanics considerations and cybernetic view points pertinent to the neurophysiological complex cohesively with the associated concourse of stochastical events and phenomena. The neurological system is a complex domain where interactive episodes are inevitable. The physics of interaction, therefore, dominates and is encountered in the random entity of neuronal microcosm. Further, the question of symmetry (or asymmetry?) that blends with the randomness of neural assembly is viewed in this book *vis-a-vis* the disorganizing effect of chance (of events) counteracted by the organizing influence of self-controlling neurocellular automata.

To comprehend and summarize the pertinent details, this book is written and organized in eight chapters:

The topics addressed in Chapters 1 through 3 are introductory considerations on neuronal activity and neural networks while the subsequent chapters outline the stochastical aspects with the associated mathematics, physics, and biological details concerning the neural system. The theoretical perspectives and explanatory projections presented thereof are somewhat unorthodox in that they portray newer considerations and battle with certain conventional dogma pursued hitherto in the visualization of neuronal interactions. Some novel features of this work are:

- A cohesive treatment of neural biology and physicomathematical considerations in neurostochastical perspectives.
- A critical appraisal of the interaction physics pertinent to magnetic spins, applied as an analogy of neuronal interactions; and searching for alternative interaction model(s) to represent the interactive neurocellular information traffic and entropy considerations.
- An integrated effort to apply the concepts of physics such as wave mechanics and particle dynamics for an analogous representation and modeling the neural activity.
- Viewing the complex cellular automata as a self-controlling organization representing a system of cybernetics.

- Analyzing the informatic aspects of the neurocybernetic complex.

This book is intended as a supplement and as a self-study guide to those who have the desire to understand the physical reasonings behind neurocellular activities and pursue advanced research in theoretical modeling of neuronal activity and neural network architecture. This book could be adopted for a graduate level course on neural network modeling with an introductory course on the neural network as the prerequisite.

If the reader wishes to communicate with the authors, he/she may send the communication to the publishers, who will forward it to the authors.

Boca Raton   P.S. Neelakanta
1994   D. De Groff

ACKNOWLEDGMENTS
The authors wish to express their appreciation to their colleagues Dr. R. Sudhakar, Dr. V. Aalo, and Dr. F. Medina at Florida Atlantic University and Dr. F. Wahid of University of Central Florida for their valuable suggestions and constructive criticism towards the development of this book.

CHAPTER 1

Introduction

1.1 General

The interconnected biological neurons and the network of their artificial counterparts have been modeled in physioanatomical perspectives, largely *via* cognitive considerations and in terms of physical reasonings based on *statistical mechanics* of interacting units. The overall objective of this book is to present a cohesive and comprehensive compendium elaborating the considerations of statistical mechanics and *cybernetic* principles in modeling real (biological) neurons as well as neuromimetic *artificial* networks. While the perspectives of statistical mechanics on neural modeling address the physics of interactions associated with the collective behavior of neurons, the cybernetic considerations describe the science of optimal control over complex neural processes. The purpose of this book is, therefore, to highlight the common intersection of statistical mechanics and cybernetics with the universe of the neural complex in terms of associated stochastical attributions.

In the state-of-the-art data-processing systems, neuromimetic networks have gained limited popularity largely due to the fragmentary knowledge of neurological systems which has consistently impeded the realistic mathematical modeling of the associated cybernetics. Notwithstanding the fact that modern information processing hinges on halfway adoption of biological perspectives on neurons, the concordant high-level and intelligent processing endeavors are stretched through the self-organizing architecture of real neurons. Such architectures are hierarchically structured on the basis of interconnection networks which represent the inherent aspects of neuronal interactions.

In order to sway from this pseudo-parasitical attitude, notionally dependent but practically untied to biological realities, the true and total revolution warranted in the *application-based artificial neurons* is to develop a one-to-one correspondence between artificial and biological networks. Such a handshake would "smear" the mimicking artificial system with the wealth of *complex automata*, the associated interaction physics, and the cybernetics of the biological neurons — in terms of information processing mechanisms with unlimited capabilities.

For decades, lack of in-depth knowledge on biological neurons and the nervous system has inhibited the growth of developing artificial networks in the image of real neurons. More impediments have stemmed from inadequate and/or superficial physicomathematical descriptions of biological systems undermining their total capabilities — only to be dubbed as totally insufficient for the requirements of advances in modern information processing strategies.

1

However, if the real neurons and artificial networks are viewed through common perspectives *via* physics of interaction and principles of cybernetics, perhaps the superficial wedlock between the biological considerations and artificial information processing could be harmonized through a binding matrimony with an ultimate goal of realizing a new generation of *massively parallel information processing systems*.

This book is organized to elucidate all those strands and strings of biological intricacies and suggest the physicomathematical modeling of neural activities in the framework of statistical mechanics and cybernetic principles. Newer perspectives are projected for the conception of better artificial neural networks more akin to biological systems. In Section 1.2, a broad outline on the state-of-the-art aspects of interaction physics and stochastical perspectives of the neural system is presented. A review on the relevant implications in the information processing is outlined. Section 1.3 introduces the fundamental considerations in depicting the real (or the artificial) neural network *via* cybernetic principles; and the basics of control and self-control organization inherent to the neural system are indicated. The commonness of various sciences including statistical mechanics and cybernetics in relation to complex neural functions is elucidated in Section 1.4; and concluding remarks are furnished in Section 1.5.

1.2 Stochastical Aspects and Physics of Neural Activity

The physics of neuronal activity, the proliferation of communication across the interconnected neurons, the mathematical modeling of neuronal assembly, and the physioanatomical aspects of neurocellular parts have been the topics of inquisitive research and in-depth studies over the past few decades. The cohesiveness of biological and physical attributions of neurons has been considered in the underlying research to elucidate a meaningful model that portrays not only the mechanism of physiochemical activities in the neurons, but also the information-theoretic aspects of neuronal communication. With the advent of developments such as the electron microscope, microelectrodes, and other signal-processing strategies, it has been facilitated in modern times to study in detail the infrastructure of neurons and the associated (metabolic) physiochemical activities manifesting as measurable electrical signals which proliferate across the interconnected neural assembly.

The dynamics of neural activity and communication/signal-flow considerations together with the associated *memory* attributions have led to the emergence of so-called *artificial neurons* and development of *neural networks* in the art of computational methods.

Whether it is the "real neuron" or its "artificial" version, the basis of its behavior has been depicted mathematically on a core-criterion that the neurons (real or artificial) represent a system of *interconnected units* embodied in a *random* fashion. Therefore, the associated characterizations depict stochastical

2

variates in the sample-space of neural assembly. That is, the neural network depicts inherently a set of implemented local constraints as connection strengths in a stochastical network. The stochastical attributes in a biological neural complex also stem from the fact that neurons may sometimes spontaneously become active without external stimulus or if the synaptic excitation does not exceed the activation threshold. This phenomenon is not just a thermal effect, but may be due to random emission of neurotransmitters at the synapses.

Further, the activities of such interconnected units closely resemble similar physical entities such as atoms and molecules in condensed matter. Therefore, it has been a natural choice to model neurons as if emulating the characteristics analogous to those of interacting atoms and/or molecules; and several researchers have hence logically pursued the statistical mechanics considerations in predicting the neurocellular statistics. Such studies broadly refer to the stochastical aspects of the collective response and the statistically unified activities of neurons viewed in the perspectives of different algorithmic models; each time it has been attempted to present certain newer considerations in such modeling strategies, refining the existing heuristics and portraying better insights into the collective activities *via* appropriate stochastical descriptions of the neuronal activity.

The subject of stochastical attributions to neuronal sample-space has been researched historically in two perspectives, namely, characterizing the response of a single (isolated) neuron and analyzing the behavior of a set of interconnected neurons. The central theme of research that has been pursued in depicting the single neuron in a statistical framework refers to the characteristics of *spike generation* (such as interspike interval distribution) in neurons. Significantly, relevant studies enclave the topics on temporal firing patterns analyzed in terms of stochastical system considerations such as random walk theory. For example, Gerstein and Mandelbrot [2] applied the *random walk models* for the spike activity of a single neuron; and modal analysis of renewal models for the spontaneous single neuron discharges were advocated by Feinberg and Hochman [3]. Further considered in the literature are the markovian attributes of the spike trains [4] and the application of time-series process and power spectral analysis to neuronal spikes [5]. Pertinent to the complexity of neural activity, accurate modeling of a single neuron stochastics has not, however, emerged yet; and continued efforts are still on the floor of research in this intriguing area despite of a number of interesting publications which have surfaced to date. The vast and scattered literature on stochastic models of spontaneous activity in single neurons has been fairly comprised as a set of lecture notes by Sampath and Srinivasan [6].

The statistics of *all-or-none* (dichotomous) firing characteristics of a single neuron have been studied as logical random bistable considerations. McCulloch and Pitts in 1943 [7] pointed out an interesting *isomorphism* between the input-output relations of idealized (two-state) neurons and the

3

truth functions of symbolic logic. Relevant analytical aspects have also since then been used profusely in the stochastical considerations of interconnected networks.

While the stochastical perspectives of an isolated neuron formed a class of research by itself, the randomly connected networks containing an arbitrary number of neurons have been studied as a distinct class of scientific investigations with the main objective of elucidating information flow across the neuronal assembly. Hence, the randomness or the *entropical* aspects of activities in the interconnected neurons and the "self-re-exciting firing activities" emulating the memory aspects of the neuronal assembly have provided a scope to consider the neuronal communication as prospective research avenues [8]; and to date the information-theoretic memory considerations and, more broadly, the neural computation analogy have stemmed as the bases for a comprehensive and expanding horizon for an intense research. In all these approaches there is, however, one common denominator, namely, the stochastical attributes with probabilistic considerations forming the basis for any meaningful analytical modeling and mathematical depictions of neuronal dynamics. That is, the global electrical activity in the neuron (or in the interconnected neurons) is considered essentially as a *stochastical process* .

More intriguingly, the interaction of the neurons (in the statistical sample space) corresponds vastly to the complicated dynamic interactions perceived in molecular or atomic ensembles. Therefore, an offshoot research on neuronal assembly had emerged historically to identify and correlate on a one-to-one basis the collective response of neurons against the physical characteristics of interacting molecules and/or atoms. In other words, the concepts of classical and statistical mechanics; the associated principles of thermodynamics; and the global functions such as the *Lagrangian*, the *Hamiltonian*, the *total entropy*, the *action*, and the *entropy* have also become the theoretical tools in the science of neural activity and neural networks. Thus from the times of Wiener [9], Gabor [10], and Griffith [11-14] to the current date, a host of publications has appeared in the relevant literature; however, there are many incomplete strategies in the formulations, several unexplained idealizations, and a few analogies with inconsistencies in the global modeling of neural activities *vis-a-vis* stochastical considerations associated with the interaction physics.

Within the framework of depicting the neural assembly as a system of interconnected cells, the activities associated with the neurons can be viewed, in general, as a collective stochastical process characterized by a random proliferation of state transitions across the interconnected units. Whether the pertinent modeling of neuronal interaction(s) evolved (conventionally) as analogous to interacting *magnetic spins* is totally justifiable (if not what is the alternative approach) the question of considering the probabilistic progression of neuronal state by an analogy of *momentum flow* (in line with *particle dynamics*) or as being represented by an analog model of *wave function*, the stochastical modeling of noise-perturbed neural dynamics and informatic aspects considered in the entropy plane of neurocybernetics are the newer perspectives which can be viewed in an exploratory angle through statistical mechanics and cybernetic considerations. A streamline of relevant bases are as follows:

- A closer look at the existing analogy between networks of neurons and aggregates of interacting spins in magnetic systems. Evolution of an alternative analogy by considering the neurons as *molecular free-point dipoles* (as in *liquid crystals* of *nematic phase* with a *long-range orientational order*) to obviate any prevalent inconsistencies of magnetic spin analogy [15].

- Identifying the class of orientational anisotropy (or persistent spatial long-range order) in the neural assembly to develop a nonlinear (squashed) input-output relation for a neural cell; and application of relevant considerations in modeling a neural network with a stochastically justifiable *sigmoidal function* [16].

- Viewing the progression of state-transitions across a neuronal assembly (consisting of a large number of interconnected cells each characterized by a dichotomous potential state) as a collective random activity similar to momentum flow in particle dynamics and development of an analogous model to describe the neural network functioning [17].

- The state-transition proliferating randomly across the neural assembly being studied as an analog of wave mechanics so as to develop a wave function model depicting the neuronal activity [17].

- Considering the inevitable presence of noise in a neuron, the change of internal states of neurons being modeled *via* stochastical dynamics [18].

1.3 Neurocybernetic Concepts

Modern information processing systems are neuromimetic and becoming more and more sophisticated as their functional capabilities are directed to emulate the diversified activities of complex neural systems. Naturally, the more we urge the functions of information processing systems to follow the complexities of the inner structure enclaved by the neural system, it becomes

5

rather infeasible to realize a tangible representation of such information processing systems to mimic closely the neuronal activities.

Hence, it calls for a shift of emphasis to project qualitatively a new viewpoint, in which the main aim is to investigate the control (and self-control) aspects of the neuronal system so as to develop information processing units emulating the image of the neural systems intact, probably with all its structural subtlety and complex control and communication protocols.

The aforesaid emphasis could be realized by adopting the concept of universal nature for control of organizing a complex system (by lowering its *entropy*) by means of standard procedures. This approach was advocated by Wiener [9] as the method of *cybernetics*, which is thenceforth known as the "science of the control and communication in complex systems, be they machines or living organisms".

The cybernetic basis for modeling the neural complex is logical in that the neural structure and its activity are inherently stochastic; and the neuronal information and/or communication processing represents an activity that fights the associated randomness, thus emphasizing the idea of a "control counteracting disorganization and destruction caused by (any) diverse random factors".

The neural complex represents an entity wherein every activity is related essentially to the collection, conversion, transmission storage and retrieval of information. It represents a system in a state which allows certain functions to be carried out. It is the *state normal*, corresponding to a set of external conditions in which the system operates. Should these conditions change suddenly, the system departs from the normal state and the new conditions set forth correspond to a new normal state. The system then begs to be transferred to this new state. In the neural complex, this is achieved first by acquiring information on the new state, and second by ascertaining how the transition of the system to the new state can be carried out. Since the change in the neuronal environment is invariably random, neither the new normal state nor how to organize a transition to it is known *a priori*. The neural complex, therefore, advocates a random search. That is, the system randomly changes its parameters until it (randomly) matches the new normal state. Eventually, this matching is self-recognized as the system monitors its own behavior.

Thus, the process of random search generates the information needed to transfer the system to the new normal state. This is an information-selection protocol with the criterion to change the system behavior approaching a new normal state, wherein the system settles down and functions normally — a condition known as *homeostasis*.

The random search-related self-organization in a neural complex follows therefore the method of cybernetics. Its self-control activity is perceived through the entity of information. Further, as well known, the notion of

6

information is based on the concepts of randomness and probability; or the self-control process of cybernetics in the neural system is dialectically united with the stochastical aspects of the associated activities.

The cybernetic basis of the neural system stems from a structured logic of details as portrayed in Figure 1.1 pertinent to the central theme of cybernetics as applied to a neural assembly. It refers to a process of control and self-control primarily from the viewpoint of neuronal information — its collection, transmission, storage, and retrieval.

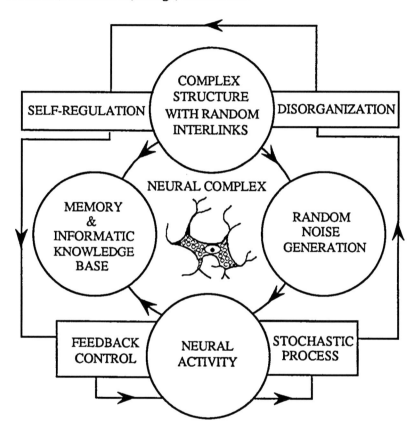

Figure 1.1 Cybernetics of neural complex

In the design of information processing systems, abstract simulation of a real (biological) neural system should comply with or mimic the cybernetic aspects depicted in Figure 1.1. Structurally, a neural complex could be modeled by a set of units communicating with each other *via* axonal links resembling the axons and dendrites of a biological neural assembly. Further, the information processing in the neural network should correspond to the self-organizing and self-adaptive (or self-behavioral monitoring) capabilities of the cybernetics associated with the biological neural complex and it activities.

The organized search process pertinent to interconnected biological neurons which enables a *dichotomous potential state* to a cellular unit corresponds to a binary threshold logic in an information processing artificial (neural) network. Classically, McCulloch and Pitts in 1943 [7] presented a computational model of nervous activity in terms of a dichotomous (binary) threshold logic. Subsequently, the process of random search in pursuit of an information selection while seeking a normal state (as governed by the self-organizing cybernetic principles) was incorporated implicitly in the artificial networks by Hebb [19]. He postulated the *principle of connectivity* (of interconnections between the cells). He surmised that the connectivity depicts a self-organizing protocol "strengthening" the pathway of connections between the neurons adaptively, confirming thereby a cybernetic mode of search procedure.

The state-transitional representation of neurons, together with the connectivity concept inculcate a computational power in the artificial neural network constituting an information processing unit. Such computational power stems from a one-to-one correspondence of the associated cybernetics in the real and artificial neurons.

In the construction of artificial neural networks two strategies are pursued. The first one refers to a *biomime*, strictly imitating the biological neural assembly. The second type is *application-based* with an architecture dictated by *ad hoc* requirements of specific applications. In many situations, such *ad hoc* versions may not replicate faithfully the neuromimetic considerations.

In essence however, both the real neural complex as well as the artificial neural network can be regarded as "machines that learn". Fortifying this dogma, Wiener observed that the concept of learning machines is applicable not only to those machines which we have made ourselves, but also is relevant to those living machines which we call animals, so that we have the possibility of throwing a new light on biological cybernetics. Further, devoting attention to those feedbacks which maintain the working level of the nervous system, Stanley-Jones [20] also considered the prospects of *kybernetic* principles as applied to the neural complex; and as rightly forecast by Wiener neurocybernetics has become a field of activity which is expected "to become much more alive in the (near) future".

The basis of cybernetics *vis-a-vis* neural complex has the following major underlain considerations:

- Neural activity stems from intracellular interactive processes.

- Stochastical aspects of a noise-infested neural complex set the associated problem "nonlinear in random theory".

- The nervous system is a memory machine with a self-organizing architecture.

- Inherent feedbacks maintain the working level of the neural complex. Essentially cybernetics includes the concept of *negative feedback* as a central feature from which the notion of *adaptive systems* and *selectively reinforced* systems are derived.

- The nervous system is a *homeostat* — it wakes up "to carry out a random search for new values for its parameters; and when it finds them, it goes to sleep again". Simply neurocybernetics depicts a search for physiological precision.

- Neural functions refer to process automation of self-control in a complex system.

- Inherent to the neural complex automata, any *symmetry constraints* on the configuration variability are unstable due to external actions triggering the (self) organization processes.

- The neural complex is a domain of information-conservation wherein the protocol of activities refers to the collection, conversion, transmission, storage, and retrieval of information.

With the enumerated characteristics as above, neurocybernetics becomes an inevitable subset of biological cybernetics — a global control and communication theory as applied to the "animal" as a whole. Therefore cybernetic attributions to the nervous system forerun their extension to the universality of biological *macrocosm*.

In the framework of cybernetics, the neural functions depicted in terms of "control and communication" activities could be expanded in a more general sense by enclaving the modern C^3I *(Command, Communication, Control, and Information)* concepts of system management .

Such an approach could address the cognitive functions involved in decision-making, planning, and control by the neural intelligence service through its synchronous, nonlinear synaptic agencies often functioning under uncertainties. Yet, it could sustain the scope of *machine-intelligence* engineering of the neural complex with the possibility of developing artificial neural networks which could mimic and pose the machine-intelligence compatible with that of real neurons.

How should neural activities be modeled *via* cybernetics? The answer to this question rests on the feasibilities of exploring the neural machine intelligence from the viewpoint of neurofunctional characteristics enumerated before.

Essentially, the neural complex equates to the cybernetics of estimating input-output relations. It is a self-organizing, "trainable-to-learn" dynamic system. It encodes (sampled) information in a framework of parallel-distributed interconnected networks with inherent feedback(s); and it is a *stochastical system*.

To portray the neural activity in the cybernetic perspectives, the following family of concepts is raised:

- The functional aspects of neurocybernetics are mediated solely by the passage of electrical impulses across neuronal cells.

- From a cybernetics point of view, neuronal activity or the neural network is rooted in mathematics and logic with a set of *decision procedures* which are typically *machine-like*.

- Neurocybernetics refers to a special class of *finite automata*,* namely, those which "learn from experience".

- In the effective model portrayal of a neuronal system, the *hardware* part of it refers to the electrical or electronic model of the neuro-organismic system involved in control and communication. Classically, it includes the models of Uttley [21], Shannon [22], Walter [23], Ashby [24], and several others. The computer (digital or analog) hardware simulation of the effective models of neurocybernetics could also be classified as a hardware approach involving a *universal machine* programmed to simulate the neural complex.

- The cybernetic approach of neural networks yields *effective models* — the models in which "if a theory is stated in symbols or mathematics, then it should be tantamount to a blueprint from which hardware could always be constructed" [25].

- The *software* aspect of effective models includes algorithms, computer programs; finite automata; information theory and its allied stochastical considerations; statistical physics and

---

* *Finite Automata* : These are well-defined systems capable of being in only a finite number of possible states constructed according to certain rules.

10

thermodynamics; and specific mathematical tools such as game theory, decision theory, boolean algebra, etc.

- The neurocybernetic complex is a richly interconnected system which has inherent self-organizing characteristics in the sense that "the system changes its basic structure as a function of its experience and environment".

- The *cognitive* faculties of neurocybernetics are *learning* and *perception*. The functional weights on each neuron change with time in such a way as to "learn". This is *learning through experience* which yields the *perceptive* attribution to the cognitive process. What is learned through past activity is *memorized*. This *reinforcement* of learned information is a *storage* or *memory* feature inherent to neurocybernetic systems.

- Homeostasis considerations of cybernetics in the self-organization procedure are applied through random search or selection of information from a noise-infested environment as perceived in a neural complex.

- The entropical and information-theoretic aspects of the neural complex are evaluated in terms of cybernetic principles.

1.4 Statistical Mechanics-Cybernetics-Neural Complex

Though the considerations of statistical mechanics and cybernetic principles as applied to neural networks superficially appear to be disjointed, there is however, a union in their applicability — it is the stochastical consideration associated with the interacting neurons. The magnetic-spin analogy based on statistical mechanics models the interacting neurons and such interactions are governed by the principles of statistics (as in magnetic spin interactions). When considering the optimal control strategies involved in self-organizing neurocybernetic processes, the statistics of the associated randomness (being counteracted by the control strategies) plays a dominant role.

Further, in both perspectives of statistical mechanics as well as cybernetics, the concepts of entropy and energy relations govern the pertinent processes involved. In view of these facts, the intersecting subsets of the neural complex are illustrated in Figures 1.2 and 1.3.

It is evident that the fields that intersect with the global neural complex functions have cross-linked attributes manifesting as unions in the Venn plane. Pertinent to such unions, the vital roots of models which have been developed for the purpose of signifying the functions of real and/or artificial neurons are embodiments of mathematics, biology (or physioanatomy), physics, engineering, and computer and informational sciences. This book delves in to the generalities of the above faculties of science, but largely adheres to statistical mechanics which deals with global properties of a large number of interacting units and cybernetics which are concerned with complex systems with constrained control efforts in seeking a self-regulation on system disorganization.

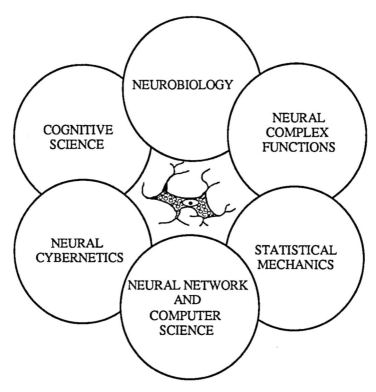

Figure 1.2 Overlaps of neural complex-related sciences

The reasons for the bifaceted (statistical mechanics and cybernetics) perspectives adopted in this book for neural network modeling stem from the sparse treatment in the literature in portraying the relevant physical concepts (especially those of cybernetics) in describing the neural network complexities. Further, the state-of-the-art informatic considerations on neural networks refer mostly to the memory and information processing relation between the neural inputs and the output; but little has been studied on the information or entropy relations pertinent to the controlling endeavors of neural self-regulation. An attempt is therefore made (in the last chapter) to present the salient aspects of informatic assays in the neurocybernetic perspectives. Collectively, the theoretical analyses furnished are to affirm the capability of the neural networks and indicate certain reliable bases for modeling the performance of neural complexes under conditions infested with intra- or extracellular perturbations on the state-transitions across the interconnected neurons.

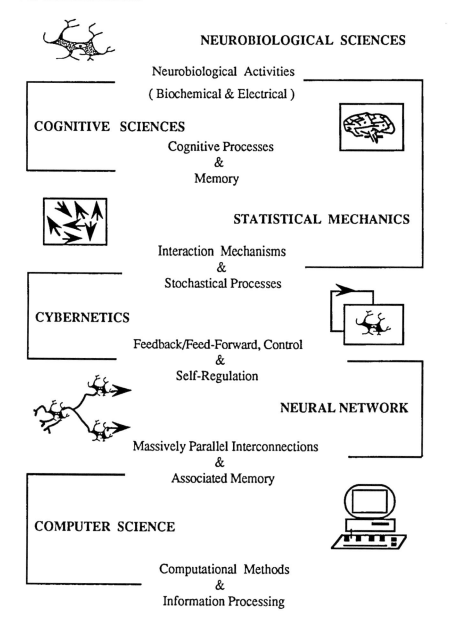

Figure 1.3 Common bases of neural theory-related sciences

1.5 Concluding Remarks

The strength of physical modeling of a neural complex lies in a coherent approach that accounts for both stochastical considerations pertinent to interacting cells and self-regulatory features of neurocybernetics. The mediating process common to both considerations is the entropy or the informational entity associated with the memory, computational, and self-controlling efforts in the neural architecture. This book attempts to address the missing links between the aforesaid considerations, the broad characteristics of which are outlined in this chapter.

CHAPTER 2

Neural and Brain Complex

2.1 Introduction

There is a hierarchy of structure in the nervous system with an inherent C^3I protocol* stemming from the brain and converging to a cell. Any single physiological action perceived (such as pain or pleasure) is the output response of a *collective activity* due to innumerable neurons participating in the decision-making control procedures in the nervous system. If the neural complex is "dubbed as a democracy, the neural activity refers to how the votes are collected and how the result of the vote is communicated as a command to all the nerve cells involved". That is, the antithesis must remain that our brain is a democracy of ten thousand million cells, yet it provides us a unified experience [14].

Functionally, the neural complex "can be regarded as a three-stage system" as illustrated in Figure 2.1. Pertinent to this three-stage hierarchy of the neural complex Griffith poses a question: "How is the level of control organized?" It is the control which is required for the C^3I protocol of neurocybernetics. A simple possibility of this is to assume the hierarchy to converge to a single cell, a dictator for the whole nervous system. However this is purely hypothetical. Such a dictatorship is overridden by the "democratic" aspect of every cell participating collectively in the decision-process of yielding an ultimate response to the (neural complex) environment.

The collectiveness of neural participation in the cybernetics of control and communication processes involved is a direct consequence of the anatomical cohesiveness of structured neural and brain complex and the associated *order* in the associated physiological activities.

The cybernetic concepts as applied to the neural complex are applicable at all levels of its anatomy — brain to cytoblast, the nucleus of the cell. They refer to the control mechanism conceivable at every neurophysiological activity — at the microscopic cellular level or at gross extensiveness of the entire brain.

In the universe of neural ensemble, the morphological aspects facilitating the neural complex as a self-organizing structure are enumerated by Kohonen [27] as follows:

- Synergic (interaction) response of compact neuronal assembly.

---

* C^3I : Command, Communication, Control, and Information — a modern management protocol in strategic operations.

- Balance (or unbalance) of inhibitory and excitatory neuronal population.

- Dimensional and configurational aspects of the ensemble.

- Probabilistic (or stochastical) aspects of individual neurons in the interaction process culminating as the collective response of the system.

- Dependence of ensemble parameters on the type of stimuli and upon the functional state of the systems.

The gross anatomical features and systematic physiological activities involved in the brain-neuron complex permit the self-organizing cybernetics in the neural system on the basis of the aforesaid functional manifestations of neuronal functions. The following sections provide basic descriptions on the anatomical and physical aspects of the neural complex.

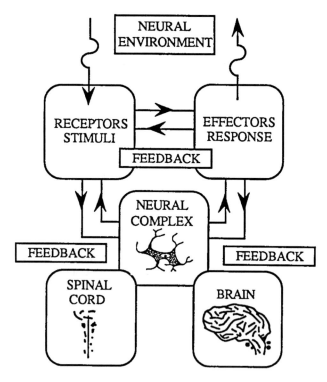

Figure 2.1 Three-stage model of a neural complex
(Adapted from [26])

2.2 Gross Features of the Brain and the Nervous System

The measured data concerning activities of the nervous system are rather limited due to its extremely complex structure with an interwining of innumerable cellular units. There are approximately 10^{10} nerve cells with perhaps 10^{14} or more interconnections in human neural anatomy which essentially consists of the *central nervous system* constituted by the brain and the spinal cord. Most of the neuronal processes are contained in these two parts excluding those which go to muscles or carry the signals from sensory organs.

The *cerebrum* is the upper, main part of the brain of vertebrate animals and consists of two equal hemispheres (left and right). In human beings, this is the largest part of the brain and is believed to control the conscious and voluntary processes. The *cerebellum* is the section of the brain behind and below the cerebrum. It consists of two lateral lobes and a middle lobe and functions as the coordinating center for the muscular movements. Highly developed cerebral hemispheres are features of *primates* and, among these, especially of human beings. It is widely surmised that this is the reason for the unique efficiency with which human beings can think abstractly as well as symbolically. Another possible reason for intellectual predominance in humans is that the brain as a whole is much more developed than the spinal cord. Even another theory suggests that such a prominence is due to the possible connection with the surface-to-volume ratio of the brain.

The brain in comparision with man-made computers is more *robust* and *fault-tolerent*. Regardless of the daily degeneration of cells, brain's physiological functions remain fairly invariant. The brain also molds itself to environment *via* learning from experience. Its information-processing is amazingly consistent even with fuzzy, random, and conjectural data. The operation of the brain is highly parallel processing and negligibly energy consuming.

The central nervous system at the microscopic level consists of [14]:

Nerve cells (or *neurons*): These are in the order of 10^{10} and are responsible for conducting the neural signaling elements. Their properties are described in the next section.

Glial cells (or *neuroglia* or *glia*): The human brain contains about 10^{11} glial cells. Unlike nerve cells, their density in different parts varies, and they fill in the spaces between the nerve cells. There has been research evidence that glial cells actually carry out functions such as memory. However, such posited glial functions are ignored invariably in neural complex modeling.

Blood vessels: These carry the blood which contains many nutrients and energy-giving materials. The main arteries and veins lie outside the central nervous system with smaller branches penetrating inwards.

Cerebrospinal fluid: This is the clear liquid surrounding the brain and spinal cord and filling the cavities (natural hollow spaces) of the brain. This liquid is the blood filtered of its white and red corpuscles and contains very little protein.

2.3 Neurons and Their Characteristics

Neurons are the building blocks of the signaling unit in the nervous system. Nerve cells come in different shapes, sizes, connections, and excitabilities. Therefore, the impression of uniformity of character which is often given for the cells is a vast oversimplification in almost all cases. However, certain properties such as *excitability*, development of an *action potential*, and *synaptic linkage* are considered as general characteristics of all nerve cells, and mathematical models of neurons are constructed based on these general features.

Each of the nerve cells has a nucleus and presumably a DNA. They do not normally divide in adult life, but they do die; and an old person may perhaps have only a third of the number of neurons at the time of birth.

Almost all outputs of the brain through neuronal transmission culminate in muscular activity. Thus, *motoneurons* — the neurons that signal the muscle fibers to contract — are deployed most frequently in the neuronal activities.

A sketch of a motoneuron is shown in Figure 2.2. It consists of three parts: The center is known as the cell-body or otherwise known as the *soma* (about 70 μm across in dimension). The cell body manufactures complex molecules to sustain the neuron and regulates many other activities within the cell such as the management of the energy and metabolism. This is the central-processing element of the neural complex.

Referring to Figure 2.2, the hair-like branched processes at the top of the cell emanating from them are called *dendrites* (about 1 mm or longer). Most input signals from other neurons enter the cell by way of these dendrites; and that leading from the body of the neuron is called the *axon* which eventually arborizes into strands and substrands as nerve fibers. There is usually only one axon per cell, and it may be very short or very long. For nerve cells (other than motoneurons where most branches go to muscle fibers), the axons terminate on other nerve cells. That is, the output signal goes down the axon to its terminal branches traveling approximately 1–100 meters/sec. The axon is the output element and it leads to a neighboring neuron. It may or may not be *myelinated*, that is, covered with a sheath of *myelin*. An axon in simple terms is a cylindrical semipermeable membrane containing *axoplasm* and surrounded by extracellular fluid.

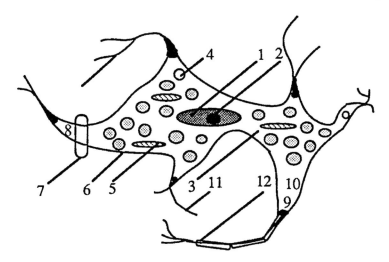

Figure 2.2 The biological neuron
1. Nucleus; 2. Nucleolus; 3. Soma; 4. Nissl body; 5. Ribosome; 6. Cell membrane; 7. Synaptic region; 8. Incoming axon; 9. Outgoing axon; 10. Axon hill; 11. Dendrite; 12. Axon sheath

The connection of a neuron's axonic nerve fiber to the soma or dendrite of another neuron is called a *synapse*. That is, the axon splits into a fine arborization, each branch of which finally terminates in a little *end-bulb* almost touching the dendrites of a neuron. Such a place of near-contact is a synapse. The synapse is a highly specialized surface that forms the common boundary between the *presynaptic* and the *postsynaptic* membrane. It covers as little as 30 nanometers, and it is this distance that a neurotransmitter must cross in the standard synaptic interaction. There are usually between 1,000 to 10,000 synapses on each neuron.

As discussed in the next section, the axon is the neuron's output channel and conveys the action potential of the neural cell (along nerve fibers) to synaptic connections with other neurons. The dendrites have synaptic connections on them which receive signals from other neurons. That is, the dendrites act as a neuron's input receptors for signals coming from other neurons and channel the postsynaptic or input potentials to the neuron's soma, which acts as an accumulator/amplifier.

The agglomeration of neurons in the human nervous system, especially in the brain, is a complex entity with a diverse nature of constituent units and mutual interconnections. The neurons exist in different types distinguished by size and degree of arborization, length of axons, and other physioanatomical details — except for the fact that the functional attributes, or principle of operation, of all neurons remain the same. The cerebellar cortex, for example, has different types of neuron multiplexed through interconnections constituting a layered cortical structure. The cooperative functioning of these neurons is essentially responsible for the complex cognitive neural tasks.

The neural interconnections either diverge or converge. That is, neurons of the cerebral cortex receive a converging input from an average of 1000 synapses and are delivered through the branching outlets to hundreds of other neurons. There are specific cells known as *Purkinje cells* in the cerebellar cortex which receive in excess of 75,000 synaptic inputs; and there also exists a single *granule cell* that connects to 50 or more Purkinje cells.

2.4 Biochemical and Electrical Activities in Neurons

The following is a description of the traditional view(s) on synaptic transmission: A very thin cell membrane separates the intracellular and extracellular regions of a biological cell shown in Figure 2.3. A high sodium and high chloride ion concentration but a low potassium concentration are found in the extracellular region, while high potassium but low sodium and low chloride concentrations are found in the intracellular region. The cellular membrane maintains this imbalance in composition through active ion transport. That is, a membrane protein, called the *sodium pump*, continuously passes sodium out of the cell and potassium into the cell. A neuron may have millions of such pumps, moving hundreds of millions of ions in and out of the cell each second. In addition, there are a large number of permanently open *potassium channels* (proteins that pass potassium ions readily into the cell, but inhibit passage of sodium). The combination of these two mechanisms is responsible for creating and maintaining the dynamic chemical equilibrium that constitutes the *resting state of the neuron* .

Under these resting conditions (steady state), one can ignore the sodium since the permeability of the biological membrane is relatively high for potassium and chloride, and low for sodium. In this case, positively-charged potassium ions (K^+) tend to leak outside the cell (due to the membrane's permeability to potassium) and the diffusion is balanced by an inward electric field that arises from the movement of these positive charges. The result is an *intracellular resting potential* of about -100 mV relative to the outside. When the cell is stimulated (due to synaptic inputs), the membrane permeability changes so that the sodium permeability greatly exceeds that of potassium and chloride. The sodium then becomes the dominant factor in establishing the steady state which arises when the inward diffusion of sodium (Na^+) elicits a counterbalancing outward electric field (and the intracellular potential becomes positive by 40 mV).

20

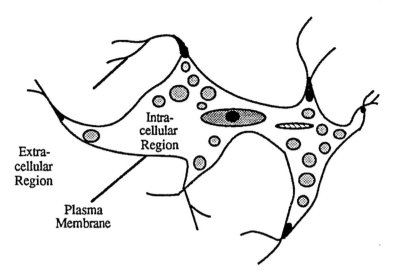

Figure 2.3 Extracellular and intracellular spaces of a
biological cell

Examining the process in greater detail, as conceived by Hodgkin and
Huxley [28], the cell *fires* (or produces an *action potential*) when
neurotransmitter molecules from the synapse reduce the potential to
approximately −50 mv. At −50 mv, voltage-controlled *sodium channels* are
opened; and sodium flows into the cell, reducing the potential even more. As
a result, further increase in sodium flow occurs into the cell and, this process
propagates to adjacent regions, turning the local cell potential to positive as it
travels. This polarity reversal spreading rapidly through the cell causes the
nerve impulse to propagate down the length of the axon to its presynaptic
connections. (The cell which has provided the *knob* where the axonal
branches end at the synapse is referred to as the *presynaptic cell*.) When the
impulse arrives at the terminal of an axon, voltage-controlled calcium
channels are opened. This causes neurotransmitter molecules to enter the
synaptic cleft and the process continues on to other neurons.

The sodium channels close shortly after opening and the potassium
channels open. As a result, potassium flows out of the cell and the internal
potential is restored to −100 mv. This rapid voltage reversal establishes the
action potential which propagates rapidly along the full length of the axon.
An electrical circuit analogy to a cell membrane can be depicted as shown in
Figure 2.4.

Action potentials refer to electrical signals that encode information by
the frequency and the duration of their transmission. They are examples of
ion movement. As the action potential travels down the axon, a large number
of ions cross the axon's membrane, affecting neighboring neurons. When

many neurons exhibit action potentials at the same time, it can give rise to relatively large currents that can produce detectable signals. Thus, neuronal transmission physically refers to a biochemical activated flow of electric signals as a collective process across the neuronal assembly.

At the end of the axon (presynapse), the electrical signal is converted into a chemical signal. The chemical signal, or neurotransmitter, is released from the neuron into a narrow (synaptic) cleft, where it diffuses to contact specialized receptor molecules embedded within the membrane of the target, or the postsynaptic neuron. If these receptors in the postsynaptic neuron are activated, channels that admit ions are opened, changing the electrical potential of the cell's membrane; and the chemical signal is then changed into an electrical signal. The postsynaptic neuron may be *excited* and send action potentials along its axon, or it may be *inhibited*. That is, the neurons are either excitatory or inhibitory (*Dale's law*). A typical cell action potential internally recorded with a microelectrode is presented in Figure 2.5.

Considering a long cylindrical axon, the neuronal propagation is nearly at a constant velocity; and the action potential can be interpreted either as a function of time at a given site or a function of position at a given time. That is, the transmembrane potential can be regarded as satisfying a *wave equation*. The stimulus intensity must reach or exceed a threshold for the neuron to fire, but the form of the action potential is not related to the exact value of stimulus intensity in the occurrence or nonoccurrence of firing activity (normally specified as the *all-or-none* response) of the cell.

2.5 Mode(s) of Communication among Neurons

As discussed earlier, a neuron is activated by the flow of chemicals across the synaptic junctions from the axons leading from other neurons. These electrical effects which reach a neuron may be *excitatory* (meaning they cause an increase in the soma potential of the receiving neuron) or *inhibitory* (meaning that they either lower the receiving neuron's soma potential or prevent it from increasing) postsynaptic potentials. If the potential gathered from all the synaptic connections exceeds a threshold value in a short period of time called the *period of latent summation*, the neuron fires and an action potential propagates down its output axon which branches and communicates with other neurons in the network through synaptic connections. After a cell fires, it cannot fire again for a short period of several milliseconds, known as the *refractory period*.

Neural activation is a *chain-like process*. A neuron is activated by other activated neurons and, in turn, activates other neurons. An action potential for an activated neuron is usually a spiked signal where the frequency is proportional to the potential of the soma. The neuron fires when the neuron's soma potential rises above some threshold value. An action potential may cause changes in the potential of interconnected neurons. The *mean firing rate* of the neuron is defined as the average frequency of the action potential. The mean soma potential with respect to the mean resting soma potential is known as the *activation level* of the neuron.

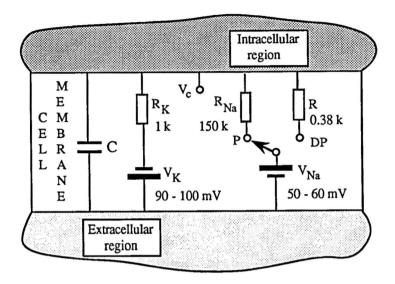

Figure 2.4 Equivalent circuit of a cell membrane

P:	Polarization; DP: Depolarization
V_c:	Intracellular potential with respect to cell exterior
V_K:	Nernst potential due to K ion differential concentration across the cell membrane
V_{Na}:	Nernst potential due to Na ion differential concentration across the cell membrane
R_K:	Relative membrane permeability to the flow of K ions
R_{Na}:	Relative membrane permeability to the flow of Na ions when the cell is polarized
R:	Relative membrane permeability to the flow of Na ions when the cell is depolarizing
C:	Capacitance of the cell

The dendrites have synaptic connections on them which receive signals from other neurons. From these synapses, the signals are then passed to the cell body where they are averaged over a short-time interval; and, if this average is sufficiently large, the cell "fires", and a pulse travels down its axon passing on to succeeding cells. Thus, the neurons relay information along structured pathways, passing messages across synapses in the traditional viewpoint, as explained above.

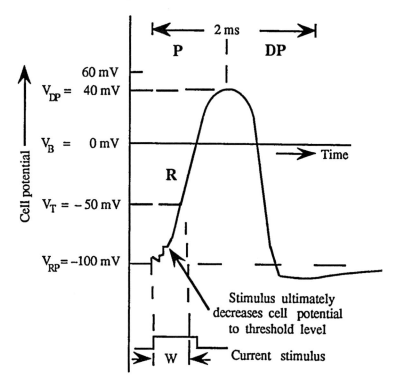

Figure 2.5 Microelectrode recording of a typical action potential

P: Polarization regime; DP: Depolarization regime; R: Regenerative breakdown regime; W: Minimum width of amplitude invariant current stimulus required for action potential generation; V_{DP}: Depolarized cell potential; V_B: Baseline potential; V_T: Threshold potential; V_{RP}: Polarized cell resting potential; Duration of spike: About 1 ms; Decay time: Up to 100 ms

Besides this classical theory, Agnati et al. [29] have advocated volume transmission as another mode of neural communication across the cellular medium constituted by the fluid-filled space between the cells of the brain; and the chemical and electrical signals travel through this space carrying messages which can be detected by any cell with an appropriate receptor. The extracellular space, which provides the fluid bathing of the neurons, occupies about 20% of the brain's volume. It is filled with ions, proteins, carbohydrates, and so on. In volume-transmission, these extracellular molecules are also regarded as participants in conveying signals. Accordingly, it has been suggested [29] that electrical currents or chemical signals may carry information *via* extracellular molecules also. Relevant electrical effects are conceived as the movement of ions (such as potassium, calcium and sodium) across the neuronal membrane. The chemical mode of volume transmission involves the release of a neuroactive substance from a neuron into the extracellular fluid where it can diffuse to other neurons. Thus, cells may communicate with each other (according to Agnati et al. [29]) without making intimate contact.

2.6 Collective Response of Neurons

The basis for mathematical modeling of neurons and their computational capabilities dwells on two considerations, namely, the associated threshold logic and the massive interconnections between them. The synergic or collective response of the neural complex is essentially stochastical due to the *randomness* of the interconnections and probabilistic character of individual neurons. A neural complex is evolved by the progressive multiplication of the interneuronal connections. As a result, the participation of an individual neuron in the collective response of the system becomes less strongly deterministic and more probabilistic. This gives rise to evoked responses of neurons being different each time to the repeated stimuli, though the reactions of the entire neuronal population (manifested as ECG, EEG, EMG, etc.) could be the same every time. Thus, the interconnected neuronal system with the stochastical input-output characteristics corresponds to a *redundant* system of *parallel connections* with a wide choice of ways for the signal propagation (or a von Neumann *random switch* — a computational basic hardware is realizable in terms of neural nets).

For the neurophysiological consideration that a neuron fires only if the total of the synapses which receive impulses in the period of latent summation exceeds the threshold, McCulloch and Pitts [7] suggested a highly simplified *computational* or *logical neuron* with the following attributes:

- A *formal neuron* (also known as the *mathematical neuron*, or *logical neuron*, or *module*) is an element with say m inputs $(x_1, x_2, ..., x_m; \quad m \geq 1)$ and one output, O where m is an axonal output or a synaptic input of a neuron. Associating weights W_i ($i \in m$) for each input and setting threshold at V_T, the module is presumed to operate at discrete time instants t_i ($i \in n$). The module "fires" or renders an output at $(n + 1)^{th}$ instant along its axon, only if the total weight of the inputs simulated at time, n exceeds V_T. Symbolically, $O(n + 1) = 1$ iff $\sum W_i x_i(n) \geq V_T$.

- The positive values of W_i (>0) correspond to excitatory synapses (that is, module inputs) whereas a negative weight $W_i < 0$ means that x_i is an inhibitory input. McCulloch and Pitts showed that the network of formal neurons in principle can perform any imaginable computation, similar to a programmable, digital computer or its mathematical abstraction, namely, the *Turing machine* [30]. Such a network has an implicit *program code* built-in *via* the coupling matrix ($\mathbf{W}_i$). The network performs the relevant computational process in parallel within each elementary unit (unlike the traditional computer wherein sequential steps of the program are executed).

- A *neural net*, or a *modular net*, is a collection of modules each operating in the same time-scale, interconnected by splitting the output of any module into a number of branches and connecting some or all of these to the inputs of the modules. An output therefore may lead to any number of inputs, but an input may come at most from one output.

- The threshold and weights of all neurons are invariant in time.

- The McCulloch-Pitts model is based on the following assumptions: Complete synchronization of all the neurons. That is, the activities of all the neurons are perceived in the same time-scale.

- Interaction between neurons (for example, the interactive electric fields across the neurons due to the associated impulses) is neglected.

- The influences of glial cell activity (if any) are ignored.

- Biochemical (hormonal, drug-induced) effects (on a short- or on long-term basis) in changing behavior of the neural complex are not considered.

In search of furthering the computational capabilities of a modular set, Hopfield in 1982 [31] queried "whether the ability of large collections of neurons to perform computational tasks may in part be a spontaneous collective consequence of having a large number of interacting simple neurons." His question has the basis that interactions among a large number of elementary components in certain physical systems yield collective phenomena. That is, there are examples occurring in physics of certain unexpected properties that are entirely due to interaction; and large assemblies of atoms with a high degree of interaction have qualitatively different properties from similar assemblies with less interaction. An example is the phenomenon of ferromagnetism (see Appendix A) which arises due to the interaction between the spins of certain electrons in the atom making up a crystal.

The collective response of neurons can also be conceived as an interacting process and has a biological basis for this surmise. The membrane potential of each neuron could be altered by changes in the membrane potential of *any* or *all* of the neighboring neurons. After contemplating data on mammals involving the average separation between the centers of neighboring cell bodies and the diameters of such cell-bodies, and finding that in the cat the dendritic processes may extend as much as 500 μm away from the cell body, Cragg and Temperley [32] advocated in favor of there being a great intermingling of dendritic processes of different cells. Likewise, the axons of brain neurons also branch out extensively. Thus, an extreme extent of close-packing of cellular bodies can be generalized in the neuronal anatomy with an intermingling of physiological processes arising thereof becoming inevitable; and, hence, the corresponding neural interaction refers to a collective activity.

It would be possible for the neurons not to interact with each neighbor, only if there were a specific circuit arrangement of fiber processes preventing such interactions. That is, an extracellular current from an active neuron would not pass through the membranes of its neighbors, if specifically dictated by an inherent arrangement. However, evidence suggests [32] that such ramification is not governed by any arrangements in some definite scheme, but rather by some random mechanical principles of growth. Therefore, each neuron seems to be forced to interact with all of its immediate neighbors as well as some more distant neurons.

The theory of a cooperative organization of neurons does not require a definite arrangement of neural processes. What is required for an assembly to be cooperative is that each of its units should interact with more than two other units, and that the degree of interaction should exceed a certain critical level. Also, each unit should be capable of existing in two or more states of different energies. This theory applies to a (statistically) large assembly of units. Considering such a cooperative assembly, a small change in the external constraints may cause a finite transition in the average properties of the whole assembly. In other words, neural interaction is an *extensive* phenomenon.

In short, neural networks are explicitly *cooperative*. The presence or absence of an action potential contributes at least two different states (*all-or-none* response), and the propagation process pertaining to dichotomous state transitions provides the mode of relevant interaction(s).

Assuming that the organization of neurons is cooperative, as mentioned earlier, there is a possible analogy between the neuronal organization and the kind of interaction that exists among atoms which leads to interatomic cooperative processes. Little [33] developed his neural network model based on this analogy. He considered the correspondence between the *magnetic Ising spin system* and the neural network. For a *magnetic spin system* which becomes ferromagnetic, the *long-range order* (defined as fixing the spin at one lattice site causes the spins at sites far away from it to show a preference for one orientation) sets in at the *Curie point* and exists at all temperatures below that critical point (see Appendix A). The onset of this long-range order is associated with the occurrence of a *degeneracy* of the maximum eigenvalue of a certain matrix describing the mathematics of the interactive process.

Considering the largest natural neural network, namely, the brain, the following mode of temporal configuration determines the long-range interaction of state-transitions: The existence of correlation between two states of the brain which are separated by a long period of time is directly analogous to the occurrence of long-range order in the corresponding spin problem; and the occurrence of these persistent states is also related to the occurrence of a degeneracy of the maximum eigenvalue of a certain matrix.

In view of the analogy between the neural network and the two-dimensional Ising problem as conceived by Little, there are two main reasons or justifications posed for such an analogy: One is that already there exists a massive theory and experimental data on the properties of ferromagnetic materials; and, therefore, it might be possible to take the relevant results and apply them to the nervous system in order to predict certain properties. The other reason is that, on account of unfamiliarity with biological aspects of the neural system, it is simple and logical to relate the relevant considerations to those with which one is already familiar.

Continuing with the same connotations, Hopfield and Tank [34] stated "that the biological system operates in a collective analog mode, with each neuron summing the inputs of hundreds or thousands of others in order to determine its graded output." Accordingly, they demonstrated the computational power and speed of collective analog networks of neurons in solving optimization problems using the principle of collective interactions.

In Hopfield's original model [31], each neuron i has two states: $\sigma_i = 0$ ("not firing") and $\sigma_i = 1$ ("firing at maximum rate"). That is, he uses essentially the McCulloch-Pitts neuron [7]. This "mathematical neuron" as deliberated earlier is capable of being excited by its inputs and of giving an output when a threshold V_T^i is exceeded. This neuron can only change its state on one of the discrete series of equally spaced times. If W_{ij} is the strength of the connection from neuron i to neuron j, a binary word of M bits consisting of the M values of σ_i represents the instantaneous state of the system; and the state progresses in time or the dynamic evolution of Hopfield's network can be specified according to the following algorithm:

$$\begin{cases} \sigma_i \to 1 & \text{if } \sum_{j \neq i} W_{ij}\sigma_j > V_T{}^i \\ \\ \sigma_i \to 0 & \text{if } \sum_{j \neq i} W_{ij}\sigma_j < V_T{}^i \end{cases} \tag{2.1}$$

Here, each neuron evaluates *randomly and asynchronously* whether it is above or below the threshold and readjusts accordingly; and the times of interrogation of each neuron are independent of the times at which other neurons are interrogated. These considerations distinguish Hopfield's net from that of McCulloch and Pitts.

Hopfield's model has stable limit points. Considering the special case of symmetric connection weights (that is, $W_{ij} = W_{ji}$), an energy functional E can be defined by a *Hamiltonian* (H_N) as :

$$H_N = E = (-1/2) \sum_{i \neq j} \sum W_{ij}\sigma_i\sigma_j \tag{2.2}$$

and ΔE due to $\Delta\sigma_j$ is given by:

$$\Delta E = -\Delta\sigma_i \sum_{j \neq i} W_{ij}\sigma_j \tag{2.3}$$

Therefore, E is a monotonically decreasing function, with the result the state-changes continue until a least local E is reached. This process is isomorphic with the Ising spin model [35]. When W_{ij} is symmetric, but has a random character (analogous to spin-glass systems where atomic spins on a row of atoms in a crystal with each atom having a spin of one half interact with the spins on the next row so that the probability of obtaining a particular configuration in the mth row is ascertained), there are known to be many (locally) stable states present as well.

In a later study, Hopfield [36] points out that real neurons have continuous input-output relations. Hence, he constructs another model based on continuous variables and responses which still retains all the significant, characteristics of the original model (based on two-state McCulloch-Pitts' threshold devices having outputs of 0 or 1 only). Hopfield let the output variable σ_i for neuron i have a squashed range $\sigma_i^0 \leq \sigma_i \leq \sigma_i^1$ and considered it as a continuous and monotone-increasing function of the instantaneous input x_i to neuron i . The typical input-output relation is then a *S-shaped sigmoid* with asymptotes σ_i^0 and σ_i^1. (The sigmoidal aspects of a neural network will be discussed in detail in a later chapter.)

2.7 Neural Net: A Self-Organizing Finite Automaton

The general characteristic of a neural net is that it is essentially a *finite automaton*. In other words, its input-output behavior corresponds to that of a finite automaton.

A *modular net* (such as the neural net) being a finite automaton has the capability for memory and computation.

Further, the modular net emerges as a computer which has command over its input and output — it can postpone its input (delay) and refer back to earlier inputs (memory) by an *effective procedure* or by a set of rules (more often known as *algorithms* in reference to computers).

A neural net in its global operation achieves a formalized procedure in deciding its input-output relation. This effective or decision procedure is typically cybernetic in that a particular operation is amenable as a mathematical operation. Further, neural net operates on a logical basis (of precise or probabilistic form) which governs the basic aspect of cybernetic principles.

A neural net supports a progression of state transitions (on-off type in the simplest neuronal configuration) — channeling a flow of bit-by-bit information across it. Thus, it envisages an information or communication protocol, deliberating the cybernetic principle.

2.8 Concluding Remarks

A neural complex has a diversified complexity in its structure and functions but portrays a unity in its collective behavior. The anatomy and physiology of the nervous system facilitates this coorperative neural performance through the mediating biochemical processes manifesting as the informational flow across the interconnected neurons. The proliferation of neural information envisages the commands, control, and communication protocols among the neurons. The resulting automaton represents a self-organizing system — a neurocybernetic.

The mechanism of interaction between the neurons immensely mimics the various interactive phenomena of statistical physics; more specifically, it corresponds to the Ising spin interaction pertinent to the statistical mechanics of ferromagnetism. Further, the neural complex is essentially a stochastical system. Its random structural considerations and conjectural functional attributes fortify such stochastical attributes and dictate a probabilistic *modus operandi* in visualizing the complex behavior of the neural system.

Modeling the biological neural complex or an artificial neural network on the basic characteristics as listed above is, therefore, supplemented by the associated stochastical theory, principles of cybernetics, and physics of statistical mechanics. The government of these considerations in essence constitutes the contents of the ensuing chapters.

CHAPTER 3

Concepts of Mathematical Neurobiology

3.1 Mathematical Neurobiology: Past and Present

In the middle of 19th century, German scientists Matthias Jakob Schleiden and Theodor Schwann proposed that all living things were made of distinct units called *cells*. They defined a cell as being a *membrane-bounded bag* containing a nucleus [29]. Neuroanatomists at that time did not realize that the brain is also made of such cells since microscopes of that era could not be used to view the brain membrane. (In fact, the membrane remained invisible until the advent of electron microscopy in the 1950's.) Many neuroanatomists believed that the entire nervous system worked as a whole independent of its individual parts. This theory has become known as the *reticular doctrine* and was advocated by the Italian anatomist Camillo Golgi. It provided a thesis that neurons communicate over relatively large distances *via* a continuous link. That is, Golgi thought that it is more likely that a neural signal is conveyed by a continuous process, rather than interrupted and somehow regenerated between the cells [29].

In 1891, the German anatomist Wilhelm Waldeyer suggested the term *neuron* and he was the first to apply cell theory to the brain. An opposing view of the reticular doctrine — the *neuron doctrine* — held that the brain was made of discrete cellular entities that only communicated with one another at specific points. The Spanish neurohistologist Santiago Ramón y Cajal amassed volumes of evidence supporting this doctrine based on microscopic techniques. This theory accounts for the view that just as electrons flow along the wires in a circuit, the neurons in the brain relay information along structured pathways. In modern notions, this concept translates into the statement that "a real neuronal network is inspired by circuit diagrams of electronics". Though polemical, in both reticular and neuronal perspectives there is, however, a "holistic conception of the brain".

Though Golgi has been criticized for his support of the reticular doctrine, there is some current evidence [29] involving volume-transmission suggesting that neural information may flow along paths that run together between relatively large, cellular territories and not only at specific points between individual cells.

A more extensive discussion on the theoretical developments concerning neuronal interactions and collective activities in actual biological systems is furnished in Chapter 5. This is based essentially on physical interaction models due to Cragg and Temperley [32] who developed the possible analogy between the organization of neurons and the kind of interaction among atoms which leads to cooperative processes in physics. Almost a decade after Cragg and Temperley projected this interaction, Griffith [13,14] tried to refute it;

3 1

however, Little's model of 1974 [33], Thompson and Gibson's model of 1981 [37], Hopfield's model of 1982 [31], and Peretto's model of 1984 [38] as well as other related models have emerged with a common perspective of viewing the neural interactions as being analogous to cooperative processes of physics as conceived by Cragg and Temperley.

Artificial neural networks, a modern trend in the art of computational science, are biologically inspired in that they perform in a manner similar to the basic functions of the biological neuron. The main advantage of using the concepts of artificial neural networks in computational strategies is that they are able to modify their behavior in response to their (input-output) environment. Both the real neurons and the artificial networks (which mimic the real neurons) have the common basis of *learning* behavior. The attribution of learning by real neurons was specified as a *learning law* by Hebb [19] as early as 1949. Hebb's rule suggests that when a cell A repeatedly and persistently participates in firing cell B, then A's efficiency in firing B is increased.

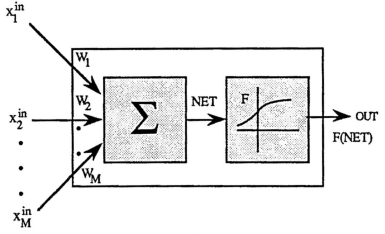

Figure 3.1 Artificial neuron

Mathematically, the degree of influence that one neuron has on another is represented by a weight associated with the interconnection between them (the biological counterpart of this interconnection is the synapse). When the neural network learns something in response to new inputs, weights are modified. That is, considering Figure 3.1, where all inputs x^{in} to a neuron are weighted by W and summed as $NET = \Sigma^{M}_{i=1} W_i x^{in}_i$, the synapse (W_{ij}) connecting neurons i and j is strengthened whenever both i and j fire. The mathematical expression which is widely accepted as the approximation of this *Hebbian rule* is given by:

$$W_{ij}(t + 1) = W_{ij}(t) + NET_i NET_j \qquad (3.1)$$

at time t. In Hebb's original model, the output of neuron i was simply NET_i. In general, $OUT_i = F(NET_i)$ where $NET_i = \Sigma_k OUT_k W_{ik}$. Most of today's training algorithms conceived in artificial neural networks are inspired by Hebb's work.

As mentioned before, McCulloch and Pitts [7] developed the first mathematical (logical) model of a neuron (see Figure 3.2). The Σ unit multiplies each input x_i^{in} by a weight W, and sums the weighted inputs. If this sum is greater than a predetermined threshold, the output is one; otherwise, it is zero. In this model, the neuron has the ability to be excited (or inhibited) by its inputs and to give an output when a threshold is exceeded. It is assumed that the neuron can only change its state at one of a discrete series of equally spaced times. In this time dependence, the logical neuron behaves differently from the actual biological one.

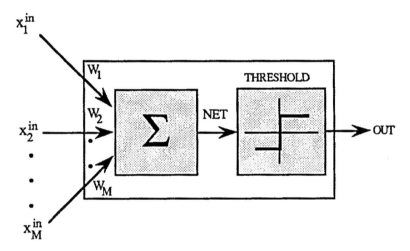

Figure 3.2 McCulloch-Pitts' model of a neuron

The McCulloch-Pitts neuron is a binary device since it exists in one of two states which can be designated as *active* and *inactive*. Hence, it is often convenient to represent its state in binary arithmetic notation, namely, it is in state 0 when it is inactive or in state 1 when it is active.

In the 1950's and 1960's, the first artificial neural networks consisting of a single layer of neurons were developed. These systems consist of a single layer of artificial neurons connected by weights to a set of inputs (as shown in Figure 3.3) and are known as *perceptrons*. As conceived by Rosenblatt [39], a simplified model of the biological mechanisms of processing of sensory information refers to *perception*. Essentially, the system receives external stimuli through the sensory units labeled as SE. Several SE units are

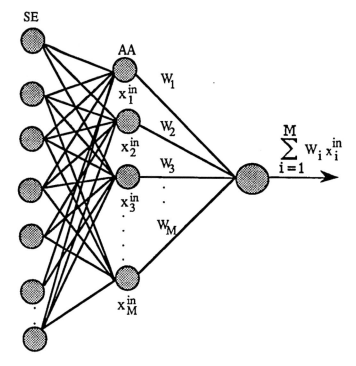

Figure 3.3 Single-layer perceptron
SE: Sensory array; AA: Associative array

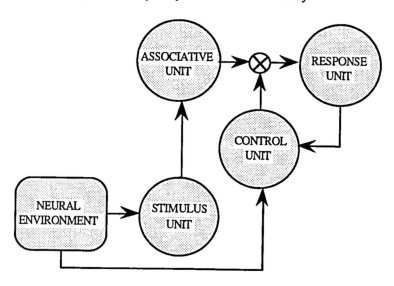

Figure 3.4 Cybernetic notions of a perceptron

connected to each associative unit (AA unit), and an AA unit is *on* only if enough SE units are activated. These AA units are the first stage or input units. As defined by Rosenblatt [40], "a perception is a network composed of stimulus-unit, association-unit, and response-unit with a variable *interactive matrix* which depends on the sequence of fast activity states of the network".

A perceptron can be represented as a logical net with cybernetic notions as shown in Figure 3.4. It was found, however, that these single-layer perceptrons have limited computational abilities and are incapable of solving even simple problems like the function performed by an exclusive-or gate. Following these observations, artificial neural networks were supposed lacking in usefulness; and hence pursuant research remained stagnant except for a few dedicated efforts due to Kohonen, Grossberg, and Anderson [41]. In the 1980's, more powerful multilayer networks which could handle problems such as the function of an exclusive-or gate, etc. emerged; and the research in neural networks has been continually growing since then.

3.2 Mathematics of Neural Activities

3.2.1 General considerations

Mathematical depiction of neural activities purport the analytical visualization of the function of real neurons. In its simplest form, as stated earlier, the mathematical neuron refers to McCulloch-Pitts' logical device, which when excited (or inhibited) by its inputs delivers an output, provided a set threshold is exceeded. An extended model improvises the time-course of the neuronal (internal) potential function describing the current values of the potential function for each neuron at an instant t, as well as at the times of firing of all attached presynaptic neurons back to the times $(t - \Delta t)$. By storing and continuously updating the potential-time data, the evolution of activity on the network as a function of time could be modeled mathematically. Thus, classically the mathematics of neurons referred to two basic considerations :

- Logical neurons.

- Time-dependent evolution of neuronal activities.

The logical neuron lends itself to analysis through boolean-space, and therefore an *isomorphism* between the bistable state of the neurons and the corresponding logic networks can be established *via* appropriate logical expressions or boolean functions as advocated by McCulloch and Pitts. Further, by representing the state of a logical network (or neurons), with a vector having 0's and 1's for its elements and by setting a threshold linearly related to this vector, the development of activity in the network can be specified in a matrix form.

The logical neuron, or McCulloch-Pitts network, has also the characteristic feature that the state vector $x(t)$ depends only on $x(t - 1)$. In other words, every state is affected only by the state at the preceding time-event. This depicts the first-order *markovian* attribute of the logical neuron model.

Further, the logical neural network follows the *principle of duality*. That is, at any time, the state of a network is given by specifying which neurons are firing at that time; or as a *duality* it would be also given, if the neurons which are not firing are specified. In other words, the neural activity can be traced by considering either the *firing-activity* or equivalently by the *nonfiring activity*, as well.

Referring to the real neurons, the action potential proliferation along the time-scale represents a time series of a state variable; and the sequence of times at which these action-potentials appear as *spikes* (corresponding to a cell firing spontaneously) does not normally occur in a regular or periodic fashion. That is, the spike-train refers to a process developing in time according to some probabilistic regime.

In its simplest form, the stochastic process of neuronal spike occurrence could be modeled as a *poissonian* process with the assumption that the probability of the cell firing in any interval of time is proportional to that time interval. In this process, the constancy of proportionality maintains that firing events in any given interval is not influenced by the preceding firing of the cell. In other words, the process is essentially regarded as *memoryless*.

The feasibility of poissonian attribution to neural activity is constrained by the condition that with the poissonian characteristic, even a single spike is sufficient to fire the cell. There is a mathematical support to this possibility on the basis of mathematical probability theory: Despite the possibility that the firing action by a neuron cell could be non-poissonian, the "pooling of a large number of non-poissonian" stochastical events leads to a resultant process which approximates to being poissonian; that is, a non-poissonian sequence of impulse train arriving at a synapse when observed postsynaptically would be perceived as a poissonian process, inasmuch as the synapses involved in this process are innumerable.

Griffith [11-14] points out that even if the process underlying the sequence of spikes is not closely poissonian, there should always be a poissonian attribute for large intervals between spikes. This is because a long time t after the cell has last fired, it must surely have lost memory exactly when it did. Therefore, the probability of firing settles down to a constant value for large t; or the time-interval distribution $p(t)$ has an exponential tail for sufficiently large t. That is, $p(t) = \lambda e^{-\lambda t}$, where λ is a constant and $\int_0^\infty p(t)\, dt = 1$. The mean time of this process $< t >$ is equal to $1/\lambda$.

The poissonian process pertinent to the neuronal spike train specifies that different interspike intervals are independent of each other. Should the neuronal spike activity be non-poissonian, deviations from spike-to-spike independence can be expected. Such deviations can be formalized in terms of the serial correlation coefficients for the observed sequence of interspike intervals. Existence of finite (nonzero) correlation coefficients may not, however, be ruled out altogether. The reasons are:

- The cellular activity is decided partly by the global activity of the brain. Therefore, depending on the part of the subject's state of activity, long-term variations in the brain-to-cell command link could possibly influence the interspike events.
- The neuronal cell is a part of an interconnected network. The multiple paths of proliferation of neuronal activity culminating at the cell could introduce correlation between the present and the future events.
- There is persistency of chemical activity in the intracellular region.

The variations in the interspike interval, if existing, render the neuronal process nonstationary which could affect the underlying probability regime not being the same at all times.

3.2.2 Random sequence of neural potential spikes

Pertinent to the neural activity across the interconnected set of cells, the probabilistic attributes of neuronal spikes can be described by the considerations of *random walk theory* as proposed by Gerstein and Mandelbrot [2]. The major objective of this model is to elucidate the probability distribution for the interspike interval with the assumption of independence for intervals and associated process being poissonian.

After the cell fires, the intracellular potential returns to its resting value (*resting potential*); and, due to the arrival of a random sequence of spikes, there is a probability p at every discrete time interval that the intracellular potential rises towards the threshold value; or there is a probability $q = (1 - p)$ of receding from the threshold potential. The discrete steps of time Δt *versus* the discrete potential change (rise or fall) Δv constitute a (discrete) random walk stochastical process. If the threshold value is limited, the random walk faces an *absorbing barrier* and is terminated. The walk could, however, be unrestricted as well, in the sense that in order to reach a threshold $v = v_0$ exactly at time t (or after t steps), the corresponding probability, p would decide the probability density of the interspike interval p_I given by [14]:

$$p_I(t,v_0) = [(v_0/8\pi pq)^{1/2}]t^{-3/2}\exp\{[v_0 - (2p - 1)t]^2/8pqt\} \qquad (3.2)$$

This refers to the probability that the interspike interval lies between t and $(t + \Delta t)$ is approximately given by $p_I \Delta t$.

Considering $f(v, t)dv$ to denote the probability at time t that the measure v of the deviation from the resting potential lies between v and $(v + dv)$, the following one-dimensional diffusion equation can be specified:

$$\partial f/\partial t = -C(\partial f/\partial v) + D(\partial^2 f/\partial v^2) \tag{3.3}$$

where C and D are constant coefficients. Gernstein and Mandelbrot used the above diffusion model equation to elucidate the interspike interval distribution (of random walk principle). The corresponding result is:

$$p_I(t, x_0) = [v_0/(4\pi D)^{1/2}]t^{-3/2}\exp[-(v - Ct)^2/4Dt] \tag{3.4}$$

Upon reaching the threshold and allowing the postsynaptic potential to decay with a time constant specified by $\exp(-\varepsilon t)$, an approximate diffusion equation for $f(v, t)$ can be written as follows :

$$\partial f/\partial t = \varepsilon \, \partial(x, f)/\partial v - C\partial f/\partial v + D\partial^2 f/\partial v^2 \tag{3.5}$$

Solution of the above equation portrays an unrestricted random passage of x to x_0 over the time and an unrestricted path of decay of the potential in the postsynaptic regime.

The classical temporal random neurocellular activity as described above can also be extended to consider the spatiotemporal spread of such activities. Relevant algorithms are based on partial differential equations akin to those of fluid mechanics. On these theories one considers the overall mean level of activity at a given point in space rather than the firing rate in any specific neuron, as discussed below.

3.2.3 Neural field theory

The spatiotemporal activity in randomly interconnected neurons refers to the neurodynamics or neural field theory in which a set of differential equations describe activity patterns in bulk neural continuum [11,12]. For example, the fluid mechanics based visualization of *neuron flow* has two perspectives: The governing differential equations could be derived on the *continuum* point of view or on the basis of a large number of interacting particles. (The later consideration refers to statistical mechanics principles which will be discussed in Chapter 5 in detail.)

The earliest *continuum model* of neuronal spatiotemporal activity is due to Beurle [42] who deduced the following set of differential equations governing the random activity in a neuronal network in terms of the level of sustained activity (F) and the proportion of cells which are nonrefractory (R):

$$d^2R/dt^2 + (1 - R\Phi/F)dR/dt = 0$$
$$dF/dt = R\Phi - F$$
$$dR/dt = -F \qquad\qquad (3.6)$$

where Φ is the probability that a sensitive cell will be energized above its threshold in unit time. The solution of the above equations represent the proliferation of the neuronal activity (in time and space) as *traveling waves*.

Considering the neuronal excitation (ψ) is "regarded as being carried by a continual shuffling between sources and fields (F_a)", Griffith in 1963 proposed [11-14] that F_a creates ψ and so on by an operation specified by:

$$H_e(x, t) = kF_a(x, t) \qquad\qquad (3.7)$$

where H_e is an undefined operation and k is a constant. The spatiotemporal distribution of the overall excitation (ψ) has hence been derived in terms of the activity of some on neurons (F_a) as:

$$\nabla^2\psi = \alpha\psi + \beta\partial\psi/dt - \gamma F_a(\psi) \qquad\qquad (3.8)$$

where α, β, γ are system coefficients.

Another continuum model of spatiotemporal activity of neurons is due to Wilson and Cowan [43,44] who described the spatiotemporal development in terms of the proportion of excitatory cells (L_e) becoming active per unit time; or proportion of inhibitory cells (L_i) becoming active per unit time. Representing the excitatory activity of the neurons by a function E_e and the inhibitory activity by E_i:

$$\mu\partial E_e/\partial t + E_e = (1 - \gamma_e E_e) D_e \qquad\qquad (3.9a)$$

$$\mu \, \partial E_i/\partial t + E_i = (1 - \gamma_i E_i) D_i \qquad\qquad (3.9b)$$

are derived as the functions to denote the spatiotemporal activity of the neurons. Here, μ, γ_e, γ_i are system coefficients and D_e, D_i are densities of excitatory and inhibitory cells participating in the regime of activity. Solutions of the above equations involve convolutions, and simplification of these equations leads to system description in terms of coupled van der Pohl oscillators.

A more involved description of the neuronal activity continuum refers to nonlinear integro-differential equations as elucidated by Orguztoreli [45] and Kawahara et al. [46]. Another modeling technique due to Ventriglia [47] incorporating intraneuron excitation, proportion of neurons in refractory state, velocity of impulses, neuronal density, synaptic density, axonic branching, and fraction of excitatory neurons in the continuum description of spatiotemporal neuronal activity has led to the study of informational waves, dynamic activities, and memory effect.

3.3 Models of Memory in Neural Networks

The memory associated with neural system is twofold: *Long-term* and *short-term memories*. The short-term memory refers to a transient activity; and if that persists long enough, it would constitute the long-term memory. The short-term memory corresponds to the input firing at a modular net stored by the impulse reverberating in the loop as illustrated in Figure 3.5. The net has a long-term memory if the short-term memory could cause its threshold to drop from 1 to 0, for example; for the memory would then be preserved and persistent even if the reverberation dies down.

The concept of memory involves a *storage mechanism* which utilizes a *storage medium*; the associated operation is termed as *memory function* which operates with the other functions of the neural network and/or the biological system. Storage and recall of information by association with other information refers to the most basic application of "collective" computation on a neural network. The information storing device is known as the *associative memory*, if it permits the recall of information on the basis of partial knowledge of its content, but without knowing its storage location. It depicts a *content addressable* memory.

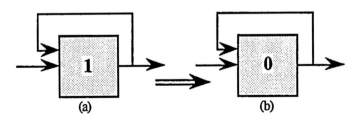

Figure 3.5 Types of memory in the neural complex
(a) Short-term memory; (b) Threshold-shift enabling long-term persistency
of state-transition

Memory models are characterized by the physical or constitutive aspects of memory functions and by the information processing abilities of the storage mechanism. The synaptic action and the state transitional considerations in the neuron (or in a set of neurons), whether transient or *persistent*, refers to a set of data or a spatiotemporal pattern of neural signals constituting an *addressable memory* on short- or long-term basis. Such a pattern could be dubbed as a *feature map*. In the interconnecting set of neurons, the associated temporal responses proliferating spatially represent a *response pattern* or a *distribution of memory*.

Relevant to this memory unit, there are *writing* and *reading* phases. The writing phase refers to the storage of a set of information data (or functionalities) to the remembered. Retrieval of this data is termed as the reading phase.

The storage of data implicitly specifies the *training* and *learning* experience gained by the network. That is, the neural network adaptively updates the synaptic weights that characterize the strength of the connections. The updating follows a set of informational training rules. That is, the actual output value is compared with a new *teacher* value; and, if there is a difference, it is minimized on *least-squares error* basis. The optimization is performed on the synaptic weights by minimizing an associated *energy function*.

The retrieval phase follows *nonlinear* strategies to retrieve the stored patterns. Mathematically, it is a single or multiple iterative process based on a set of equations of dynamics, the solution of which corresponds to a neuronal value representing the desired output to be retrieved.

The learning rules indicated before are pertinent to two strategies, *unsupervised* learning rule and *supervised* learning rule. The unsupervised version (also known as *Hebbian learning*) is such that, when unit i and j are simultaneously excited, the strength of the connection between them increases in proportion to the product of their activation. The network is trained without the aid of a teacher *via* a training set consisting of input training patterns only. The network learns to adapt based on the experiences collected through the previous training patterns.

In the supervised learning, the training data has many pairs of input/output training patterns. Figure 3.6 illustrates the supervised and unsupervised learning schemes.

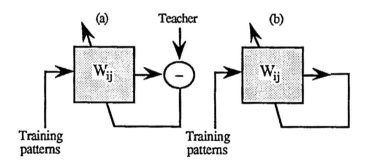

Figure 3.6 Learning schemes
(a) Supervised learning; (b) Unsupervised learning (Adapted from [48])

Networks where no learning is required are known as *fixed-weight* networks. Here the synaptic weights are *prestored*. Such an association network has one layer of *input neurons* and one layer of *output neurons*.

Pertinent to this arrangement, the pattern can be retrieved in one shot by a *feed-forward* algorithm; or the correct pattern is deduced *via* many iterations through the same network by means of a *feedback* algorithm. The feed-forward network has a linear or nonlinear associative memory wherein the synaptic weights are precomputed and prestored. The feedback associative memory networks are popularly known as *Hopfield nets*.

3.4 Net Function and Neuron Function

The connection network neurons are mathematically represented by a *basis* function U (W, x) where W is the weight matrix and x is the input matrix. In hyperplane, U is a linear basis function given by:

$$U_i (W, x) = \sum_{j=1}^{n} W_{ij} x_j \tag{3.10}$$

and in hypersphere representation the basis function is a second-order function given by:

$$U_i (W, x) = [\sum_{j=1}^{n} (x_j - W_{ij}^2)]^{1/2} \tag{3.11}$$

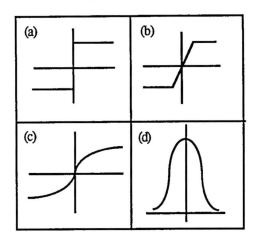

Figure 3.7 Activation functions
(a) Step function; (b) Ramp function; (c) Sigmiodal function;
(d) Gaussian function (Adapted from [48])

The net value as expressed by the basis function can be transformed to depict the nonlinear activity of the neuron. This is accomplished by a nonlinear function known as the *activation function*. Commonly, step, ramp, sigmoid, and gaussian functions are useful as activation functions. These are illustrated in Figure 3.7.

3.5 Concluding Remarks

Mathematical representation of neural activity has different avenues. It may be concerned with a single neuron activity or the collective behavior of the neural complex. Single neuron dynamics refers to the stochastical aspects of biochemical activity at the cells manifesting as trains of spikes. The collective behavior of neural units embodies the interaction between massively connected units and the associated memory, adaptive feedback or feed-forward characteristics, and the self-organizing controlling endeavors. The analytical representation of memory functions of the neural complex governs the learning (or training) abilities of the network exclusively *via* the transient and/or persistent state of the system variables.

Another mathematical consideration pertinent to the neural system refers to the spatiotemporal dynamics of the state-transitional proliferations across the interconnected neurons. Existing models portray different analogical "flow" considerations to equate them to the neuronal flow. Equations of wave motion and informational transit are examples of relevant pursuits.

The contents of this chapter as summarized above provide a brief outline on the mathematical concepts as a foundation for the chapters to follow.

CHAPTER 4

Pseudo-Thermodynamics of Neural Activity

4.1 Introduction

Randomly interconnected neurons emulate a redundant system of parallel connections with almost unlimited routings of signal proliferation through an enormous number of *von Neumann random switches*. In this configuration, the energy associated with a neuron should exceed the limit set by the thermodynamic minimum (Shannon's limit) of a logical act given by:

$$k_B T \ln(2) = 3 \times 10^{-14} \text{ erg} \tag{4.1}$$

where $k_B T$ is the Boltzmann energy. The actual energy of a neuron is about 3×10^{-3} erg per binary transition being well above the thermodynamical (noise) energy limit as specified above.

Not only the energy dissipative aspects per neuron could be delved in thermodynamic perspective (as von Neumann did), the state-transitional considerations (*flow-of-activation*) across randomly interconnected neurons constitute a process activity which can be studied under *thermodynamic principles*.

Such a thermodynamical attribution to neural activity stems from the inherent statistical characteristics associated with the spatiotemporal behavior of the neuronal nets. Since the thermodynamics and neural networks have a common intersection of statistical considerations as primary bases, Bergstrom and Nevanlinna [49] postulated in 1972 that the state of a neural system could be described by its *total neural energy* (E) and its *entropy distribution* ($\mathcal{H}$). *Entropy* here refers to the *probability* or *uncertainity* associated with the *random switching* or state-transitional behavior of the neural complex. The governing global attributes of such a description are that the total energy remains invariant (*conservation principle*) and the neural complex always strives to maximize its entropy. Therefore, the entropy of the neural system is decided by the total energy, number of neuronal cells, and number of interconnections. The principle of maximum entropy as applied to systems of interacting elements (such as the neural network) was advocated by Takatsuji in 1975 [50]; and the relevant thermodynamic principles as applied to the neural network thereof have led to the so-called *machine* concepts detailing the learning/training properties of neural nets, as described below.

As discussed earlier the neural system *learns* or *is trained* by means of some process that modifies its weights in the collective state-transitional activity of interconnected cells. If the training is successful, application of a set of inputs to the network produces the desired set of outputs. That is, an *objective function* is realized. Pertinent to real neurons, the training method follows a stochastical strategy involving random changes in the weight values of the interconnections retaining those changes that result in improvements. Essentially, the output of a neuron is therefore a weighted sum of its inputs operated upon by some nonlinear function (F) characterized by the following basic training or procedural protocol governing its state-transitional behavior:

- A set of inputs at the neuron results in computing the outputs.

- These outputs are compared with desired (or target) outputs; if a difference exists, it is measured. The measured difference between input and output in each module is squared and summed.

- The object of training is to minimize this difference known as the *objective function*.

- A weight is selected randomly and adjusted by a small random amount. Such an adjustment, if reduces the objective function, is retained. Otherwise, the weight is returned to its previous value.

- The above steps are iterated until the network is trained to a desired extent, that is, until the objective function is attained.

Basically, the training is implemented through random adjustment of weights. At first large adjustments are made, retaining only those weight changes that reduce the objective function. The average step-size is then gradually reduced until a global minimum is eventually reached. This procedure is akin to the thermodynamical process of *annealing* in metals.[*] In molten state, the atoms in a metal are in incessant random motion and therefore less inclined to reach a *minimum energy state*. With gradual cooling,

---

[*] *Annealing* in the metallurgical sense refers to the physical process of heating a solid until it melts, followed by cooling it down until it crystallizes into a state with a perfect lattice structure. During this process, the free-energy of the solid is minimized. The cooling has to be done gradually and carefully so as not to get trapped in locally optimal lattice structures (metastable states) with crystal imperfections. (Trapping into a metastable state occurs when the heated metal is *quenched* instantaneously, instead of being cooled gradually.)

however, lower and lower energy states are assumed until a *global minimum* is achieved and the material returns to a crystalline state.

Adoption of *thermodynamic annealing* to neural activity refers to achieving the global energy minimum criterion. For example, in Hopfield nets, it is possible to define an *energy function* that is monotonically decreasing; and state changes in these nets continue until a minimum is reached. However, there is no guarantee that this will be the global minimum; and it is, in fact, most likely that the minimum will be one of the many *locally stable states*. Therefore, in solving for output-input relations in such a net, the *optimum solution* may not be realized.

That is, a difficulty encountered normally with Hopfield nets is the, tendency for the system to stabilize at a *local* rather than going to a *global* minimum. This can, however, be obviated by introducing noise at the input so that the artificial neurons change their state in a statistical rather than in a deterministic fashion. To illustrate this concept, a ball rolling up and down in a terrain can be considered. The ball may settle at a local trap (L_m) such that it may not be able to climb up the global minimum valley (see Figure 4.1). However, a strategy which introduces some disturbances can cause the ball to become unsettled and jump out from the *local minima*. The ball being at L_m, corresponds to a weight setting initially to a value L_m. If the random weight steps are small, all deviations from L_m increase the objective function (energy) and will be rejected. This refers to *trapping* at a local minimum. If the weight setting is very large, both the local minimum at L_m and the global minimum at G_m are "frequently revisited"; and the changes in weight occur so drastically that the ball may never settle into a desired minimum.

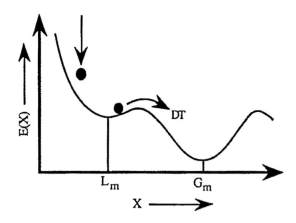

Figure 4.1 Global and local minima
G_m: Global minimum; L_m: Local minimum; X: Weight state; E(X): Objective function or cost function; DT: Escape from local minima (de-trapping) corresponds to annealing

By starting with large steps and gradually reducing the size of the average random step, the network could, however, escape from the local minima ensuring an eventual network stabilization. This process mimics the metallurgical annealing described above. This *simulated annealing* enables a combinatorial optimization of finding a solution among a potentially very large number of solutions with minimal cost-function. Here, the cost-function corresponds to the free-energy on a one-to-one basis.

The *annealing* in a network can be accomplished as follows: When a *disturbance* is deliberately introduced and started at a random state at each time-step, a new state could be generated according to a *generating probability density*. This new state would replace the old state, if the new state has a lower energy. If it has higher energy, it is designated as a new state with a probability as determined by an *acceptance function*. This way, jumps occasionally are allowed to configurations of higher energy. Otherwise the old state is retained. In the search of minimal energy solution, there are possibilities of other suboptimal solutions emerging arbitrarily close to an optimum. Therefore, reaching the optimal solution invariably warrants a rather extensive search with massive computational efforts.

4.2 Machine Representation of Neural Network

In practice, the system incorporated with a method of introducing a *disturbance* or a *random noise* for the purpose of state *de-trapping* (as explained above) is referred to as a *machine*. For example, Hinton et al. [51] proposed the *Boltzmann statistics* of thermodynamics to describe the neural system as a *machine* representing the "constant satisfaction networks that learn" by the implementation of local constraints as connection strength in stochastic networks. In these Boltzmann machines, the generating probability density is *gaussian* given by :

$$G_G(x) = \exp[-x^2/T_G{}^2(t)] \qquad (4.2)$$

where the time-schedule of changing fluctuations in the machine is described in terms of an *artifical cooling temperature* $T_G(t)$ (also known as the *pseudo-temperature*) being inversely logarithmic to time; and the *acceptance probability* (corresponding to the chance of the ball climbing a hump) follows the *Boltzmann distribution*, namely:

$$G_A = 1/\{1 + \exp[-\Delta E/T_G(t)]\} \qquad (4.3)$$

where ΔE is the increase in energy incurred by a transition. It may be noted that both the *acceptance* and *generating functions* are decided essentially by the *cooling schedule*. The above probability distribution refers to the probability distribution of energy states of the annealing thermodynamics, that is, the probability of the system being in a state with energy ΔE. At high temperatures, this probability approaches a single value for all energy states, so that a high energy state is as likely as a low energy state. As the temperature is lowered, the probability of high energy states decreases as compared to the probability of low energy states. When the temperature approaches zero, it becomes very unlikely that the system will exist in a high energy state.

The Boltzmann machine is essentially a *connectionist model* of a neural network: It has a large number of interconnected elements (neurons) with bistable states and the interconnections have real-valued strengths to impose local constraints on the states of the neural units; and, as indicated by Aarts and Korst [52], "a *consensus function* gives a quantitative measure for the 'goodness' of a global configuration of the Boltzmann machine determined by the states of all individual units".

The cooperative process across the interconnections dictates a simple, but powerful massive parallelism and distribution of the state transitional progression and hence portrays a useful configuration model. Optimality search *via* Boltzmann statistics provides a substantial reduction of computational efforts since the simulated annealing algorithm supports a massively parallel execution. Boltzmann machines also yield higher order optimizations *via* learning strategies. Further, they can accommodate *self-organization* (through learning) in line with the *cybernetics* of the human brain.

Szu and Hartley [53] in describing neural nets advocated the use of a *Cauchy machine* instead of the Boltzmann machine. The Cauchy machine uses the generating probability with the Cauchy/Lorentzian distribution given by:

$$G_C(x) = T(t)/[T_C^2(t) + x^2] \qquad (4.4)$$

where $T_C(t)$ is the pseudothermodynamic temperature. It allows the cooling schedule to vary inversely proportional to the time, rather than to the logarithmic function of time. That is, Szu and Hartley used the same acceptance probability given by Equation (4.3), but $T_G(t)$ is replaced by $T_C(t)$.

The reason for Szu and Hartley's modification of the generating probability is that it allows a fast annealing schedule. That is, the presence of a small number of very long jumps allows faster escapes from the local minima. The relevent algorithm thus converges much faster. Simulation done by Szu and Hartley shows that the Cauchy machine is better in reaching and staying in the global minimum as compared with the Boltzmann machine. Hence, they called their method the *fast simulated annealing* (FSA) schedule.

48

Another technique rooted in thermodynamics to realize annealing faster than the Cauchy method refers to adjusting the temperature reduction rate according to the *(pseudo) specific heat* calculated during the training process. The metallurgical correspondence for this strategy follows:

During annealing, metals experience *phase* changes. These phases correspond to discrete energy levels. At each phase change, there is an abrupt change of the specific heat defined as the rate of change of temperature with energy. The change in specific heat results from the system settling into one of the local energy minima. Similar to metallurgical phase changes, neural networks also pass through phase changes during training. At the phase transistional boundary, a specific heat attribution to the network can therefore be considered which undergoes an abrupt change. This *pseudo specific heat* refers to the average rate of change of pseudo-temperature with respect to the objective function. Violent initial changes make the average value of the objective function virtually independent of small changes in temperature so that the specific heat is a constant. Also, at low temperatures, the system is frozen into a minimum energy. Thus again, the specific heat is nearly invariant. As such, any rapid temperature fluctuations at the temperature extrema may not improve the objective function to any significant level.

However, at certain *critical temperatures* (such as a ball having just enough energy for a transit from L_m to G_m, but with insufficient energy for a shift from G_m to L_m), the average value of the objective function makes an abrupt change. At these critical points, the training algorithm must alter the temperature very slowly to ensure the system not trapping into a local minimum (L_m). The critical temperature is perceived by noticing an abrupt decrease in the specific heat, namely, the average rate of change with the objective function. Upon reaching the objective function, the temperature maximal to this value must be traversed slowly enough so as to achieve a convergence towards a global minimum. At other temperatures, a larger extent of temperature reduction can, however, be used freely in order to curtail the training time.

4.3 Neural Network *versus* Machine Concepts

4.3.1 Boltzmann Machine

The Boltzmann machine has a binary output characterized by a stochastic decision and follows instantaneous activation in turn. Its activation value refers to the net input specified by:

$$NET_i = \sum_i^N W_{ij} o_j + \theta_i + e_n \qquad (4.5)$$

where e_n is the error (per unit time) on the input caused by random noise, and o_j is the output value that an ith neuron receives from other neuron units through the input links such that o_j assumes a graded value over a range $0 < o_j < 1$. The neuron has also an *input bias*, θ_i. Unit $i \in N$ has state $o_i \in \{0, 1\}$ so that the global state-space S of this machine is 2^N. Associated with each state $s \in S$ is its *consensus* C_s defined as $\Sigma_{ij} W_{ij} o_j o_i$. The Boltzmann machine maximizes C_{si} within the net through the simulated annealing algorithm *via* pseudo-temperature T which asymptotically reduces to zero. For any fixed value of $T > 0$, the Boltzmann machine behaves as an irreducible *Markov chain* tending towards equilibrium. This can be explained as follows.

A finite *Markov chain* represents, in general, a sequence $o(n)$ $(n = \dots -1, 0, +1\dots)$ probability distributions over the finite state-space S. This state-space refers to a stochastic system with the state changing in discrete epochs; and $o(n)$ is the probability distribution of the state of the system in epoch, n such that $o(n + 1)$ depends *only* on $o(n)$ and not on previous states. The transition from state s to s' in the Markov chain is depicted by a transitional probability $p_{ss'}$. The Markov chain can be said to have attained an equilibrium, if the probability of the state-space $o_s(n)$ remains invariant as π_s for all s and n. π_s is referred to as the *stationary distribution;* and it is *irreducible*, if the set $\{\pi_s\}$ has nonzero cardinality.

Writing $p_{ss'} = (g_{iss'} p_{as'})$, $g_{iss'}$ is the probability of choosing i, a global choice from N, and is taken to be uniformly 1/n; whereas $p_{as'}$ is the probability of making the change once i has been chosen and is determined locally by the weight at a critical unit where s and s', being adjacent, differ. The parameter $g_{ss'}$ is the generating probability and $p_{as'}$ is the acceptance probability considered earlier. That is, for a Boltzmann machine $p_{ass'} = 1/[1 + \exp(\Delta ss')]$, with $\Delta ss' = (C_s - C_{s'})/T$. Hence $\Delta s's = \Delta ss'$ so that $p_{as's} = (1 - p_{ass'})$

In terms of the consensus function, the stationary distribution of a Boltzmann machine can be written as:

$$\pi_s = \exp[(C_s - C_{smax})/T] \tag{4.6}$$

where C_{smax} is the maximum consensus over all states. Annealing refers to $T \rightarrow 0$ in the stationary distribution π_s. This reduces π_s to a uniform distribution over all the maximum consensus states which is the reason for the Boltzmann machine being capable of performing global optimization.

If the solution of binary output is described probabilistically, the output value o_i is set to one with the probability $p_{o_i} = 1$ regardless of the current state. Further:

$$p_{o_i} = 1/[1 + \exp(-\Delta E_i/T)] \qquad (4.7)$$

where ΔE_i is the change in energy identifiable as NET_i and T is the pseudo-temperature. Given a current state i with energy E_i, then a subsequent state j is generated by applying a small disturbance to transform the current state into a next state with energy E_j. If the difference $(E_j - E_i) = \Delta E_i$ is less than or equal to zero, the state j is accepted as the current state. If $\Delta E_j > 0$, the state j is accepted with a probability as given above. This rate of probability acceptance is also known as the *Metropolis criterion*. Explicitly, this acceptance criterion determines whether j is accepted from i with a probability:

$$p_a(\text{accept } j) = \begin{cases} 1 & \text{if} & f(i) \le f(i) \\ \exp[f(i) - f(j)]/C_p & \text{if} & f(j) \ge f(i) \end{cases}$$

$$(4.8)$$

where $f(i)$, $f(j)$ are *cost-functions* (equivalent of energy of a state) in respect to solutions i and j; and C_p denotes a *control parameter* (equivalent to the role played by temperature). The Metropolis algorithm indicated above differs from the Boltzmann machine in that the transition matrix is defined in terms of some (positive) energy function E_S over S. Assuming $E_S \ge E_{s'}$:

$$p_{ss'} = 1 \qquad (4.9a)$$
$$p_{s's} = \exp[(E_{s'} - E_S)/T] \qquad (4.9b)$$

It should be noted that $p_{ss'} \ne p_{s's}$. The difference is determined by the intrinsic ordering on $\{s, s'\}$ induced by the energy function.

Following the approach due to Akiyama et al. [54], a machine can in general be specified by three *system parameters*, namely, a *reference activation level*, a_o; *pseudo-temperature*, T; and *discrete time-step*, Δt. Thus, the system parameter space for a Boltzmann machine is $S(a_o = 0, T, \Delta t = 1)$, with the output being a unit step-function. The distribution of the output o_i is specified by the following moments:

$$Mean: \quad <o_i> = \int_{a=0}^{\infty} (1) \, p(a_i = a) da = \Phi(<a_i>/\sigma_{a_i}) \qquad (4.10)$$

$$Variance: \sigma^2_{a_i} = \int_{a=0}^{\infty} (1) \, p(a_i = a) da - <o_i^2>$$

$$= \Phi(<a_i>/\sigma_{a_i}) \{1 - \Phi [<a>/\sigma_{a_i}]\} \qquad (4.11)$$

where $\Phi(x)$ is the standard cumulative gaussian distribution defined by:

$$\Phi(x) = \int_{-\infty}^{+x} \exp(-x^2/2)dx \qquad (4.12)$$

It may be noted that o_i being binary, $<o_i>$ refers to the probability of o_i equal to 1; and the cumulative gaussian distribution is a *sigmoid* which matches the probability function of the Boltzmann machine, defined by Equation (4.3).

As indicated before, the Boltzmann machine is a basic model that enables a solution to a combinational problem of finding the optimal (the best under constrained circumstances) solution among a "countably infinite" number of alternative solutions. (Example: *The traveling salesman problem*[*].)

In the *Metropolis algorithm* (as the Boltzmann's acceptance rule), if the lowering of the temperature is done adequately slowly, the network reaches thermal equilibrium at each pseudo-temperature as a result of a large number of transitions being generated at a given temperature value. This "thermal" equilibrium condition is decided by Boltzmann distribution, which as indicated earlier refers to the probability of state i with energy E_i at a pseudo-temperature T. It is given by the following conjecture:

$$p_T (x = i) = [\exp(- E_i/k_B T)]/Z(T) \qquad (4.13)$$

where $Z(T)$ is the *partition function* defined as $Z(T) = \Sigma \exp[E_j/(k_B T)]$ with the summation over all possible states. It serves as the normalization constant in Equation (4.13).

4.3.2 McCulloch-Pitts Machine

Since the McCulloch-Pitts model has a *binary output* with a *deterministic decision* and *instantaneous activation* in time, its machine parameter space can be defined by: $S_m(a_0 = 0, T = 0, \Delta t = 1)$. The corresponding output is a unit step function, assuming a totally deterministic decision.

---

[*] *The traveling salesman problem*: A salesman, starting from his headquarters, is to visit each town in a prescribed list of towns exactly once and return to the headquarters in such a way that the length of his tour is minimal.

4.3.3 Hopfield Machine

Contrary to the McCulloch-Pitts model, the Hopfield machine has a *graded output* with a *deterministic decision* but with *continuous (monotonic) activation* in time. Therefore, its machine parameter is $S_m(a_0, 0, \Delta t)$. If the system gain approaches infinity, then the machine parameter becomes $S_m(0, 0, 0)$. The neuron model employed in the generalized *delta rule*[*] is described by $S_m(a_0, 0, 1)$, since it is a discrete time version of the Hopfield machine.

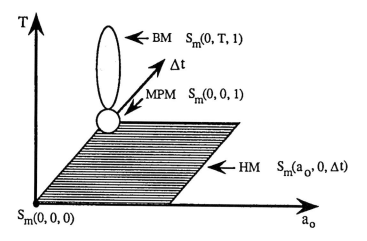

Figure 4.2 Parametric space of the machines
BM: Boltzmann machine; MPM: McCulloch-Pitts machine
HM: Hopfield machine (Adapted from [54])

---

[*] *Delta rule (Widrow-Hoff rule):* The delta rule is a training algorithm which modifies weights appropriately for target and actual outputs (of either polarity) and for both continuous and binary inputs and outputs. Symbolically denoting the correction associated with the ith input x_i by Δ_i, the difference between the target (or desired output) and the actual output by δ_0 and a learning rate coefficient by η_ℓ, the delta rule specifies Δ_i as equal to $(\eta_\ell)(\delta_0)(x_i)$. Further, if the value of ith weight after adjustment is $W_i(n + 1)$, it can be related to the value of ith weight before adjustment, namely, $W_i(n)$ by the equation $W_i(n + 1) = W_i(n) + \Delta_i$.

4.3.4 Gaussian Machine

Akiyama et al. [54] proposed a machine representation of a neuron and termed it as a *gaussian machine* which has a *graded response* like the Hopfield machine, and behaves stochastically as the Boltzmann machine. Its output is influenced by a random noise added to each input and as a result forms a probabilistic distribution. The relevant machine parameters allow the system to escape from local minima.

The properties of the gaussian machine are derived from the normal distribution of random noise added to the neural input. The machine parameters are specified by $S_m(a_0, T, \Delta t)$. The other three machines discussed earlier are special cases of the Gaussian machine as depicted by the system parameter space shown in Figure 4.2.

4.4 Simulated Annealing and Energy Function

The concept of equilibrium statistics stems from the principles of statistical physics. A basic assumption concerning many-particle systems in statistical physics refers to the *ergodicity hypothesis* in respect to *ensemble averages* which determine the average of *observed values* in the physical system at thermal equilibrium. Examples of physical quantities which can be attributed to the physical system under such thermal equilibrium conditions are *average energy*, *energy-spread* and *entropy*. Another consideration at the thermal equilibrium is *Gibbs' statement* that, if the *ensemble* is *stationary* (which is the case if equilibrium is achieved), its density is a function of the energy of the system. Another feature of interest at thermal equilibrium is that, applying *the principle of equal probability*, the probability that the system is in a state i with energy E_i, is given by Gibbs' or Boltzmann's distribution indicated earlier. In the annealing procedure, also detailed earlier, the probabilities of global states are determined by their energy levels. In the search of global minimum, the stability of a network can be ensured by associating an *energy function*[*] which culminates in a minimum value. Designating this energy function as the *Lyapunov function*, it can be represented in a recurrent network as follows:

$$E = (-1/2) \sum_i \sum_j W_{ij} o_i o_j - \sum_j x_j o_j + \sum_j V_{Tj} o_j \qquad (4.14)$$

where E is an *artifical network energy function* (Lyapunov function), W_{ij} is the weight from the output of neuron i to the input of neuron j, o_j is the output of neuron j, x_j is the external input to neuron j, and V_{Tj} represents

---

[*] The term *energy function* is derived from a physical analogy to the magnetic system as discussed in Appendix A.

the threshold of neuron j. The corresponding change in the energy ΔE due to a change in the state of neuron j is given by:

$$\Delta E = -[\underset{i \neq j}{\Sigma}(W_{ij}o_i) + x_j - V_{T_j}](\Delta o_j)$$

$$= -[NET_j - V_{T_j}](\Delta o_j) \qquad (4.15)$$

where Δo_j is the change in the output of neuron, j.

The above relation assures that the network energy must either decrease or stay invariant as the system evolves according to its dynamic rule regardless of the net value being larger or less than the threshold value. When the net value is equal to V_T, the energy remains unchanged. In other words, any change in the state of a neuron will either reduce the energy or maintain its present value. The continuous decreasing trend of E should eventually allow it to settle at a minimum value ensuring the stability of the network as discussed before.

4.5 Cooling Schedules

These refer to a set of parameters that govern the convergence of simulated annealing algorithms. The cooling schedule specifies a finite sequence of values of the control parameters (Cp) involving the following steps:

- An intial value C_{P_0} (or equivalently, an initial temperature T_0) is prescribed.

- A decrement function indicating the manner in which the value of the control parameter decreases is specified.

- The final value of the control parameter is stipulated as per a *stop criterion.*

A cooling schedule is also in conformity of specifying a finite number of transitions at each value of the control parameter. This condition equates to the simulated annealing algorithms being realized by generating homogeneous chains of finite length for a finite sequence of descending values of the control parameter.

A general class of cooling schedule refers to a *polynomial-time cooling schedule.* It leads to a polynomial-time execution of the simulated algorithm, but it does not guarantee the deviation in cost between the final solution obtained by the algorithm and the optimal cost.

The Boltzmann machine follows a simple annealing schedule with a probability of a change in its objective function, as decided by Equation (4.2).

The corresponding scheduling warrants that the rate of temperature reduction is proportional to the reciprocal of the logarithm of time to achieve a convergence towards a global minimum. Thus, the *cooling rate* in a Boltzmann machine is given by [55] :

$$T(t) = T_0/\log(1 + t) \qquad (4.16)$$

where T_0 is the initial (pseudo) temperature and t is the time. The above relation implies almost an impractical cooling rate, or the Boltzmann machine often takes an infinitely large time to train.

The Cauchy distribution is *long tailed* which corresponds to increased probability of large step-sizes in the search procedure for a global minimum. Hence, the Cauchy machine has a reduced training time with a schedule given by:

$$T(t) = T_0/(1 + t) \qquad (4.17)$$

The simulated annealing pertinent to a gaussian machine has a hyperbolic scheduling, namely,

$$T(t) = T_0/(1 + t/\tau_T) \qquad (4.18)$$

where τ_T is the time-constant of the annealing schedule.

The initial value of the control parameter (T_0), in general, should not be large enough to allow virtually all transitions to be accepted. This is achieved by having the *initial acceptance ratio* χ_0 (defined as the ratio of initial number of accepted transitions to the number of proposed transitions) close to unity. This corresponds to starting with a small T_0 multiplied by a constant factor greater than 1, until the corresponding value of χ_0 calculated from generated transitions approaches 1. In metallurgical annealing, this refers to heating up the solid until all particles are randomly arranged in the liquid phase.

The functional decrement of the control parameter (T) is chosen so that only small changes in control parameters would result. The final value of the control parameter (T) corresponds to the termination of the execution of the algorithm when the cost function of the solution obtained in the last trial remains unchanged for a number of consecutive chains with a Markov structure. The length of the Markov chain is bounded by a finite value compatible to the small decremental value of the control parameter adopted.

In the network optimization problems, the change in the reference level with time adaptively for the purpose of a better search is termed as *sharpening schedule*. That is, sharpening refers to altering the output gain curve by slowly decreasing the value of the reference activation level (a_0) over the time-scale. The candidates for the sharpening scheme are commonly exponential, inverse-logarithm, or linear expressions. For gaussian machines, a hyperbolic sharpening schedule of the type:

$$a_0 = A_0/(1 + t/\tau_{ao}) \tag{4.19}$$

has been suggested. Here A_0 is the initial value of a_0 and τ_{ao} is the time constant of the sharpening schedule.

In general, the major problems that confront the simulated annealing is the convergence speed. For real applications, in order to guarantee fast convergence, Jeong and Park [56] developed lower bounds of annealing schedules for Boltzmann and Cauchy machines by mathematically describing the annealing algorithms *via* Markov chains. Accordingly, the simulated annealing is defined as a Markov chain consisting of a transition probability matrix P(k) and an annealing schedule T(k) controlling P for each trial, k.

4.6 Reverse-Cross and Cross Entropy Concepts

In all the network trainings, it can be observed that simulated annealing is a stochastic strategy of searching the ground state by minimizing the energy or the *cooling function*. Pertinent to the Boltzmann machine Ackley et al. [57] projected alternatively a learning theory of minimizing the *reversed-cross entropy* or the *cross-entropy functions*, as briefed below:

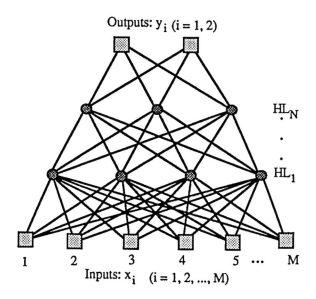

Figure 4.3 A multilayered perceptron with hidden layers $HL_1, ..., HL_N$

A typical neural net architecture is structured macroscopically as *layers* or *rows* of units which are fully interconnected as depicted in Figure 4.3. Each unit is an information processing element. The first layer is a fanout of processing elements intended to receive the inputs x_i and distribute them to the next layer of units. The hierarchical architecture permits that each unit in each layer receives the output signal of each of the units of the row (layer) below it. This continues until the final row which delivers the network's estimate o' of the correct output vector o. Except for the first row which receives the inputs and the final row which produces the estimate o', the intermediate rows or layers consist of units which are designated as the *hidden layers*.

Denoting the probability of the vector state of *visible neurons* (units) as $P'(v_\alpha)$ under the free-running conditions (with the network having no environmental input), and the corresponding probability determined by the environment as $P(V_\alpha)$, a *distance parameter* can be specified as an objective function for the purpose of minimization. Ackley et al. [57] employed reverse cross-entropy (RCE) as defined below to depict this distance function:

$$G_{RCE} = \sum_\alpha P(V_\alpha) \ln[P(V_\alpha)/P'(V_\alpha)] \qquad (4.20)$$

The machine adjusts its weight W_{ij} to minimize the distance G_{RCE}. That is, it seeks a negative gradient of the derivative $(\partial G_{RCE}/\partial W_{ij})$ *via* an estimate of this derivative. In reference to the Boltzmann machine, this gradient is specified by:

$$(\partial G_{RCE}/\partial W_{ij}) = (p_{ij} - p'_{ij})/T \qquad (4.21)$$

where p_{ij} is the average probability of two units (i and j) both being the on-state when the environment is clamping the states of the visible neurons, and p'_{ij} is the corresponding probability when the environmental input is absent and the network is free-running on its own internal mechanism as a cybernetic system. To minimize G_{RCE}, it is therefore sufficient to observe (or estimate) p_{ij} and p'_{ij} under thermal equilibrium and to change each weight by an amount proportional to the difference between these two quantities. That is:

$$\Delta W_{ij} \propto (p_{ij} - p'_{ij}) \qquad (4.22)$$

Instead of the reverse cross-entropy (RCE), a cross-entropy parameter (G_{CE}) as defined below has also been advocated by Liou and Lin [58] as an alternative strategy for the aforesaid purposes:

$$G_{CE} = \sum_\alpha P'(V_\alpha) \ln[P'(V_\alpha)/P(V_\alpha)] \qquad (4.23)$$

4.7 Activation Rule

In the neural network, the relation between the net input (NET_i) and its output value (o_j) is written in a simple form as in Equation (4.5). When the neuron is activated by the input (NET_i), the *activation value* (a_i) of the neuron is altered (with respect to time) by a relation written as :

$$(\Delta a_i/\Delta t) = -(a_i/\tau) + NET_i \qquad (4.24)$$

where τ is the time-constant of the neuronal activation. By specifying the reference activation level as a_o, the output value o_j of the neuron can be determined by the graded response of the neuron. Written in a functional form:

$$o_j = F(a_i/a_o) \qquad (4.25)$$

where F is a monotonic function which limits the output value between upper and lower bounds. It is, therefore, a *squashing function* which is S-shaped or *sigmoidal.* The reference activation level a_o is termed as the *gain factor.* It is the *first system parameter,* and the random error e_n is a noise term whose variance is dictated by the (pseudo) temperature which can be regarded as the *second system parameter.*

4.8 Entropy at Equilibrium

Relevant to the combinational optimization problem, the simulated annealing algorithm specified by the conjecture, namely, the distribution (Equation 4.8), p_{ai} refers to a stationary or equilibrium distribution which guarantees asymptotic convergence towards globally optimal solutions. The corresponding *entropy at equilibrium* is defined as:

$$\mathcal{H}_i(T) = -\sum p_{ai}(T) \ln[P_{ai}(T)] \qquad (4.26)$$

which is a natural *measure of disorder.* High entropy corresponds to *chaos* and low entropy values to *order.* Pertinent to the neural net, entropy also measures the *degree of optimality.* Associated energy of the state i, namely, E_i with an acceptance probability p_{ai} has an expected valued $<E_i>_T$ which refers to the *expected cost* at equilibrium. By general definition through the first moment:

$$\langle E_i \rangle_T = \sum_i E_i(P_{ai}) \tag{4.27}$$

Likewise, the second moment defines the expected square cost at equilibrium. That is:

$$\langle E_i^2 \rangle = \sum_i E_i^2(P_{ai}) \tag{4.28}$$

and a variance of the cost can be specified as :

$$\sigma_i^2 = [\langle E_i^2 \rangle] - [\langle E_i \rangle]^2 \tag{4.29}$$

Considering the neural complex as a large physical ensemble, from the corresponding principles of statistical thermodynamics, the following relations can be stipulated:

$$\partial \langle E_i \rangle / \partial T = \sigma_i^2 / T^2 \tag{4.30}$$

and

$$\partial \mathcal{H}_i / \partial T = \sigma_i^2 / T^3 \tag{4.31}$$

These conditions indicate that in simulated annealing, the expected cost and entropy decreases monotonically — provided equilibrium is reached at each value of the control parameter (T) to their final value, namely, E_{iopt} and $\ln |\mathcal{H}_{iopt}|$, respectively.

Further, the entropy function $\mathcal{H}_i$ under limiting cases of T are to be specified as follows:

$$\underset{T \to \infty}{\text{Limit}} \; \mathcal{H}_i(T) = \ln |S| \qquad i \in S \tag{4.32a}$$

and

$$\underset{T \to \infty}{\text{Limit}} \; \mathcal{H}_i(T) = \ln |S_{opt}| \qquad i \in S_{opt} \tag{4.32b}$$

where S and S_{opt} are the sets of the states and globally optimal states, respectively. (In the combinational problems S and S_{opt} denote the sets of solutions and globally optimal solutions, respectively.)

In statistical physics, corresponding to the ground state, $S_o = \log(1) = 0$ defines the *third law of thermodynamics*.

When the annealing algorithm refers to the equilibrium distribution, the probability of finding an optimal solution (or state) increases monotonically with decreasing T. Further, for each suboptimal solution there exists a positive value of the pseudo-temperature T_i (or the control parameter), such that for $T < T_i$ the probability of finding that solution decreases monotonically with decreasing T. That is,

$$
\partial p_{ai}/\partial t \quad
\begin{aligned}
&> 0 && \text{if } T < T_i \\
&= 0 && \text{if } T = T_i \\
&< 0 && \text{if } T > T_i
\end{aligned}
\qquad (4.33)
$$

4.9 Boltzmann Machine as a Connectionist Model

Apart from the neural networks being modeled to represent the neurological fidelity, the adjunct consideration in the relevant modeling is the depiction of neural computation. Neural computational capability stems predominantly from the massive connections with variable strengths between the neural (processing) cells. That is, a neural complex or the *neuromorphic* network is essentially a *connectionist model* as defined by Feldman and Ballard [59]. It is a massively distributed parallel-processing arrangement. Its relevant attributes can be summarized by the following architectural aspects and activity-related functions:

- Dense interconnection prevails between neuronal units with variable strengths.
- Strength of interconnections specifies degree of interaction between the units.
- The state of the connectionist processing unit has dichotomous values (corresponding to firing or non-firing states of real neurons). That is, $o_i \in \{0, 1\}$.
- The neural interaction can be inhibitory or excitatory. The algebraic sign of interconnecting weights depicts the two conditions.
- The response that a unit delegates to its neighbor can be specified by a scalar nonlinear function (F) with appropriate connection strength. That is:

$$
o_i = F \left[\sum_{j \in N} W_{ij}\, o_j \right] \qquad (4.34)
$$

where o_i is the response of unit i, W_{ij} is the strength of interconnection and N represents the set of neighbors of i.
- Every unit operates in parallel simultaneously changing its state to the states of its neighbors. The dynamics of states lead to the units settling at a

steady (nonvarying) value. The network then freezes at a global configuration.

- The units in the network cooperatively optimize the global entity of the network with the information drawn from the local environmant.

- The network information is thus distributed over the network and stored as interconnection weights.

- The Boltzmann machine (or a connectionist model) has only dichotomous states, $o_i \in \{0, 1\}$. In contrast to this, neural modeling has also been done with continuous valued states as in the Hopfield and Tank [34] model pertinent to neural decision network.

- In the Boltzmann machine, the response function F is stochastic. (There are other models such as the *perceptron model*, where the response function F is regarded as deterministic.)

- The Boltzmann machine is a symmetrical network. That is, its connections are bidirectional with $W_{ij} = W_{ji}$. (Models such as a feed-forward *network*, however, assume only a unidirectional connections for the progression of state transitional information.)

- The Boltzmann machine is adaptable for both *supervised* and *unsupervised training*. That is, it can "learn" by capturing randomness in the stimuli they receive from the environment and adjust their weights accordingly (unsupervised learning), or it can also learn from a set of classification flags to learn to identify the correct output.

- The Bolzmann machine represents a model of hidden layers of units which are not visible in the participation of neural processing, and these hidden units capture the higher order disturbances in the learning process.

Construction of a Boltzmann machine relies on the following considerations as spelled out by Aarts and Korst [52]: "The strength of a connection in a Boltzmann machine can be considered as a quantitative measure of the desirability that the units joined by the connection are both 'on'. The units in a Boltzmann machine try to reach a maximal *consensus* about their individual states, subject to the desirabilities expressed by the connection strengths. To adjust the states of the individual units to the states of their neighbors, a probabilistic state transition mechanism is used which is governed by the simulated annealing algorithm."

4.10 Pseudo-Thermodynamic Perspectives of Learning Process

Neural models represent general purpose learning systems that begin with no initial object-oriented knowledge. In such models, *learning* refers to incremental changes of probability that neurons are activated.

Boltzmann machines as mentioned before have two classes of learning capabilities. They can learn from *observations*, without supervision. That is, the machine captures the irregularities in its environment and adjusts its internal representation accordingly. Alternatively, the machines can learn from examples and counterexamples of one or more concepts and induce a general description of these concepts. This is also known as *supervised learning*. The machine that follows unsupervised learning is useful as *content addressable memory*. The learning capabilities of Boltzmann machines are typical of connectionist network models.

Invariably some of the units in a Boltzmann machine are clamped to a specific state as dictated by the environment. This leaves the machine to adjust the states of the remaining units so as to generate an output that corresponds to the most probable interpretation of the incoming stimuli. By this, the network acquires most probable environmental configuration, with some of its environmental units fixed or clamped.

The environment manifests itself as a certain probability distribution by interacting with the Boltzmann machine *via* a set $v_u \subset N$ of visible (external units) while the remaining units $h_u \subset N$ are hidden and are purely internal. The visible units are clamped to states by samples of environment imposed on them. Under this connection, a learning algorithm permits the determination of appropriate connection weights so that the hidden units can change, repeated by over a number of learning cycles in which the weights are adjusted. The degree of such adjustment is determined by the behavior of the machine under the *clamped mode* as compared to the *normal (free-running) mode*.

Pertinent to the clamped mode, Livesey [60] observes that such a mode is not an intrinsic characteristic of the learning algorithm associated with the Boltzmann machine, but rather a condition stipulated by the transition probabilities of Markov chain depicting the state-transitional stochastics of these machines. The relevant condition refers to the underlying *time reversibility* under equilibrium conditions. A machine in equilibrium is *time reversible* when it is not possible to tell from its state which way time is flowing. In other words, the chain and its time reversal are identical. This happens under a *detailed balanced* condition given by:

$$\pi_s p_{ss'} = \pi_t p_{s's} \qquad s', s \in S \qquad (4.35)$$

The essence of machine representation of a neural network embodies a *training procedure* with an algorithm which compares (for a given set of network inputs) the output set with a desired (or a target) set, and computes the error or the difference [61]. For a given set of synaptic coupling $\{W_{ij}\}$, denoting the training error in terms of the energy function by $\xi(\{W_{ij}\})$, this ensemble can be specified *via* the equilibrium statistical mechanics concept by *Gibbs' ensemble* with the distribution function specified by $\exp[-(\xi_S(\{W_{ij}\})/k_BT]$, where k_BT represents the (pseudo) Boltzmann energy.

Here $x(\{W_{ij}\})$ is pertinent to a subsystem which is taken as the representative of the total collection of subsystems of the total neuronal ensemble. Considering the *partitioning* of the weights among the energy states given by $x(\{W_{ij}\})$, the following Gibbs' relation can be written in terms of a *partition function* :

$$p_M(\{W_{ij}\}) = p_0(\{W_{ij}\})\exp[-\xi(\{W_{ij}\})/k_BT]/Z_M$$

$$(4.36)$$

where p_0 is existing probability distribution imposing normalization constraints on the system parameters, $p_M(\{W_{ij}\})$ is the Gibbs' distribution pertinent to the M associated trainings (or a set of M training examples), and Z_M is the partition function defined as:

$$Z_M = \int_{-\infty}^{+\infty} p_0(\{W_{ij}\})\exp[-\beta\xi(\{W_{ij}\})] \prod_{j=1}^{N} dW_{ij}$$

$$(4.37)$$

where $\beta = 1/k_BT$, and N is the total number of couplings.

The average training error per example (e_{tr}) can be specified by the (pseudo) thermodynamic (Gibbs') *free - energy*, G defined as :

$$G = -(\beta N)^{-1} <\ln Z_M>_{En}$$

$$(4.38)$$

where $<...>_{En}$ is the average over the ensemble of the training examples. That is:

$$e_{tr} = <<\xi(\{W_{ij}\})>_T>_{En}/M = (\partial G/\partial \beta)(N/M)$$

$$(4.39)$$

64

where $<...>_T$ is the thermal average. The above relation implies that the free-energy (and hence the training error) are functions of the relative number of training examples and the Boltzmann energy.

From the free-energy relation, the corresponding thermodynamic entropy can be deduced *via* conventional *Legendre transformation*, given by:

$$\mathcal{H} = \beta(Ne_{tr}/M - G)$$
$$= <<\ln[P_0(\{W_{ij}\})/P_M(\{W_{ij}\})]>_T>_{En}$$

(4.40)

This entropy function is a measure of the deviation of P_M from the initial distribution P_0. At the onset of training, $(M/N) = 0$. Therefore $\mathcal{H} = 0$. As the training proceeds, $\mathcal{H}$ becomes negative. The entropy measure thus describes the evolution of the distribution in the system parameter space.

Akin to the molal free-energy (or the *chemical potential**) of thermodynamics, the corresponding factor associated with the relative number of training examples is given by:

$$\mu = \partial G/\partial(M/N)$$
$$= (e_{tr} - \mathcal{H}_0/\beta)$$

(4.41)

where $\mathcal{H}_0$ is the *one-step entropy* defined as :

$$\mathcal{H}_0 = << \ln[P_{M-1}(\{W_{ij}\})/P_M(\{W_{ij}\})] >_T>_{En}$$

(4.42)

The one-step entropy is a measure (specified by a small or a large negative number) to describe qualitatively the last learning step resulting in a small or large contraction of the relevant subspace volume respectively.

4.11 Learning from Examples Generated by a Perceptron

The *perceptron*, in general, refers to a system with an input-output relation dictated by a nonlinear squashing function. The learning rule of a test perceptron corresponds to a target to be learned as in a reference perceptron with $\{x'_j\}$ inputs, $\{W'_{ij}\}$ coupling weights resulting in the output o'_j. The corresponding sets for the test perceptron are taken as $\{x_j\}$ and $\{W_{ij}\}$. Because x_j and x'_j are not identical, a correlation coefficient can be specified in terms of a joint gaussian distribution of the variates as indicated below :

---

* *Chemical potential* : It is the rate of change of free-energy per mole (in a chemical system) at constant volume and temperature.

$$\rho(x_j; x'_j) = (1/2\pi\sigma_x) \exp[-(x_j - x'_j)^2/2\sigma_x^2 - (x_j^2/2)] \qquad (4.43)$$

where σ_x^2 is the variance of the inputs.

Since the output is binary, the associated error measure $e_o(x)$ is also binary. That is, $e_o(x) = 0$ or 1, if $x < 0$ or $x > 0$, respectively. The corresponding total training error is therefore:

$$x_M(\{W_{ij}\}) = \sum_{k=1}^{M} e_o(-x_k x'_k) \qquad (4.44)$$

which permits an explicit determination of the partition function (Z_M) defined by Equation (4.37).

The ensemble average of the partition function can also be represented in terms of an average of a logarithm, converted to that of a power as follows:

$$<\ln Z_M>_{En} = (d <Z^n_M>_{En}/dn)|_{n=0}$$
$$= -\beta NG \qquad (4.45)$$

In the case of completely random examples, $<Z^n_M>_{En}$ bifurcates into two parts, one being a power of the connectivity N and the other that of the number M of the training examples. Further, in reference to the correlation coefficients (Equation 4.43), the averaging process leads to two quantities which characterize the ensemble of interconnected cells. They are:

1. *Overlap parameter* (R^a) pertinent to the reference perceptron defined as :

$$R^a = (1/N) \sum_{j=1}^{N} W_j^a W_j^r \qquad (4.46)$$

2. *Edwards-Anderson parameter* which specifies the overlap between *replicas* given by:

$$q^{ab} = (1/N) \sum_{j=1}^{N} W_j^a W_j^b - \delta^{ab} \qquad (4.47)$$

where δ^{ab} is the Kronecker delta.

In the dynamics of neural network training, the basic problem is to find the weighting parameters W_{ij} for which a set of configuration (or patterns) $\{\xi_i^\mu\}$ ($i = 1, 2, 3, ..., N$; $\mu = 1, 2, 3, ..., p$) are stationary (fixed) points of the dynamics. There are two lines of approach to this problem. In the first approach, the W_{ij} are given a specific storage (or memory) prescription. The so-called Hebb's rule which is the basis of Hopfield's model essentially follows this approach. Another example of this strategy is the *pseudo-inverse rule* due to Kohonen which has been applied to Hopfield's net by Personnaz et al. [62] and studied analytically by Kanter and Sompolinsky [63]. Essentially, W_{ij} are assumed as *symmetric* (that is, $W_{ij} = W_{ji}$) in these cases.

In the event of a mixed population of symmetric and asymmetric weights, *asymmetry parameters* η_s can be defined as follows:

$$\eta_s = \sum_{i \neq j}^{N} W_{ij} W_{ji} / \sum_{i=j}^{N} W^2_{ij} \qquad (4.48)$$

or equivalently:

$$\eta_s = \sum_{i \neq j} \left[(W_{ij}^{sy})^2 - (W_{ji}^{asy})^2 \right] / \sum_{i \neq j} (W_{ij})^2 \qquad (4.49)$$

where $W^{sy,asy} = 1/2(W_{ij} \pm W_{ji})$ are the symmetric and asymmetric components of W_{ij}, respectively. When $\eta_s = 1$, the matrix is fully symmetric and when $\eta_s = -1$, it is fully asymmetric. When $\eta_s = 0$, W_{ij} and W_{ji} are fully correlated on the energy, implying that the symmetric and asymmetric components have equal weights.[*]

As mentioned earlier, the network training involves finding a set of stationary points which affirm the convergence towards the target configuration or pattern. The two entities whose stationary values are relevant to the above purpose are the *overlap parameters*, namely, q and R. The stationary values of q and R can be obtained *via replica symmetry ansatz* solution (see Appendix C).

4.12 Learning at Zero Temperature

This refers to a naive training strategy of minimizing the (training) error. It specifies a critical relative number of learning examples below which training with zero error is possible and beyond which, however, error in the training process cannot be avoided. The error normally arises from external noise in the examples. Absence of such noise permits a perfect learning process, and the target rule can be represented by the reference perceptron. The

---

[*] Another common measure of symmetry is the parameter k_s defined by $W_{ij} = W_{ij}^{sy} + k_s W_{ij}^{asy}$ and is related to η_s by $\eta_s = (1 - k_s^2)/(1 + k_s^2)$.

criticality can therefore be quantified as a function of the *overlap parameters* R. Thermodynamically, an excess of the number of learning examples (beyond the critical value) makes the system unstable.

The entities q and R represent *order parameters* which vary as a function of the relative number of training examples for a given generic noise input. The R parameter exhibits a nonmonotonic behavior and is the *precursor of criticality*. When R → 1, the training approaches the pure reference system, despite of the presence of noise; or, the system self-regulates as a *cybernetic complex* and organizes itself so that the learning process filters out the external noise. The convergence of R towards 1 obviously depends on the amount of noise introduced. The smaller the external noise is the faster the convergence is. Criticality is the limit of capacity for error-free learning in the sense that the critical number of training examples brings about a *singularity* in the learning process, as it is indicated by the behavior of the training error and the different examples. Further, the criticality marks the onset of *replica symmetry breaking*, implying that the parameter space of interaction with minimal training error breaks up into disconnected subsets.

The *naive learning* strategy discussed earlier minimizes the training error for a given number of examples. It also results in a *generalization error*. That is, in characterizing the achievement in learning *via* R (representing a deviation from the reference perceptron), the probability that the trained perceptron makes an error in predicting a noise output of the reference perceptron is also implicitly assessed; and an error on an example independent of the training set, namely:

$$x^{av} = <<<e_0(-x_i x'_i)>_{E_n}'>_{E_n}>_T \qquad (4.50)$$

could be generalized. (Here, the prime refers to the new example.)

Explicit evaluation of this generalization error indicates that it decreases monotonically with R. In other words, a maximal overlap of R is equivalent to minimizing the generalization error. Hence the algebraic consequence of R → 0 translates into algebraic decay of the generalization error, and such a decay is slower if the examples contain external noise. However, by including the thermal noise into the learning process, the system acquires a new degree of freedom and allows the minimization of the generalization error as a function of temperature. Therefore, the naive learning method (with T → 0) is not an optimal one. In terms of the (functional) number of examples (M/N), the effect of introducing thermal synaptic noise, with a noise parameter (ϑ_N), a threshold curve M/N (ϑ_N) exists such that for M/N < (>) M/N(ϑ_N) the optimum training temperature is zero (positive).

4.13 Concluding Remarks

In summary, the following could be considered as the set of (pseudo) thermodynamic concepts involved in neural network modeling:

- Thermodynamics of learning machines.
- Probability distributions of neural state transitional energy states.
- Cooling schedules, annealing, and cooling rate.
- Boltzmann energy and Boltzmann temperature.
- Reverse-cross and cross entropy concepts.
- System (state) parameters.
- Equilibrium statistics.
- Ensemble of energy functions (Gibbs' ensemble).
- Partition function concepts.
- Gibbs' free-energy function.
- Entropy.
- Overlaps of replicas.
- Replica symmetry ansatz.
- Order parameters.
- Criticality parameter.
- Replica symmetry breaking.
- Concept of zero temperature.

Evolution of the aforesaid concepts of (pseudo) thermodynamics and principles of statistical physics as applied to neural activity can be summarized by considering the chronological contributions from the genesis of the topic to its present state as stated below. A descriptive portrayal of these contributions is presented in the next chapter.

- McCulloch and Pitts (1943) described the neuron as a binary, all-or-none element and showed the ability of such elements to perform logical computations [7].
- Gabor (1946) proposed a strategy of finding solutions to problems of sensory perception through quantum mechanic concepts [10].
- Wiener (1948) suggested the flexibility of describing the global properties of materials as well as "rich and complicated" systems *via* principles of statistical mechanics [9].
- Hebb (1949) developed a notion that a percept or a concept can be represented in the brain by a cell-assembly with the

suggestion that the process of learning is the modification of synaptic efficacies [19].

- Cragg and Temperley (1954) indicated an analogy between the persistent activity in the neural network and the collective states of coupled magnetic dipoles [32].

- Caianiello (1961) built the neural statistical theory on the basis of statistical mechanics concepts and pondered over Hebb's learning theory [64].

- Griffith (1966) posed a criticism that the Hamiltonian of the neural assembly is totally unlike the ferromagnetic Hamiltonian [13].

- Cowan (1968) described the statistical mechanics of nervous nets [65].

- Bergstrom and Nevalinna (1972) described a neural system by its total neural energy and its entropy distribution [49].

- Little (1974) elucidated the analogy between noise and (pseudo) temperature in a neural assembly thereby paving "half the way towards thermodynamics" [33].

- Amari (1974) proposed a method of statistical neurodynamics [66].

- Thompson and Gibson (1981) advocated a general definition of *long range order* pertinent to the proliferation of the neuronal state transtitional process [37].

- Ingber (1982,1983) studied the statistical mechanics of neurocortical interactions and developed dynamics of synaptic modifications in neural networks [67,68].

- Hopfield (1982,1984) completed the linkage between thermodynamics *vis-a-vis* spin glass in terms of models of content-addressable memory through the concepts of entropy and provided an insight into the energy functional concept [31,36].

- Hinton, Sejnowski, and Ackley (1984) developed the *Boltzmann machine* concept representing "constraint satisfaction networks that learn" [51].

- Peretto (1984) searched for an *extensive quantity* to depict the Hopfield-type networks and constructed formulations *via stochastic units* which depict McCulloch and Pitts weighted-sum computational neurons but with the associated dynamics making "mistakes" with a certain probability analogous to the temperature in statistical mechanics [38].

- Amit, Gutfreund, and Sompolinsky (1985) developed pertinent studies yielding results on a class of stochastic network models such as Hopfield's net being amenable to exact treatment [69].

- Toulouse, Dehaene, and Changeux (1986) considered a spin-glass model of learning by selection in a neural network [70].
- Rumelhart, Hinton and Williams (1986) (re)discovered *back-propagation* algorithm to match the adjustment of weights connecting units in successive layers of multilayer perceptions [71].
- Gardner (1987) explored systematically the space of couplings through the principles of statistical mechanics with a consequence of such strategies being applied exhaustively to neural networks [72].
- Szu and Hartley (1987) adopted the principles of thermodynamic *annealing* to achieve energy minimum criterion, and proposed the Cauchy machine representation of neural network [53].
- 1989: The unifying concepts of neural networks and spin glasses were considered in the collection of papers presented in the Stat. Phys. 17 workshop [73].
- Aarts and Korst (1989) elaborated the stochastical approach to combinational optimization and neural computing [52].
- Akiyama, Yamashita, Kajiura and Aiso (1990) formalized *gaussian machine* representation of neuronal activity with graded response like the Hopfield machine and stochastical characteristics akin to the Boltzmann machine [54].

71

CHAPTER 5

The Physics of Neural Activity:
A Statistical Mechanics Perspective

5.1 Introduction

On neural activity, several of the earliest theoretical papers appearing from the 1940's into the 1970's [74] dealt mostly with random network models of interconnected cells. Typically considered were the dynamics of the overall activity level of neurocellular random networks and the relation of dynamic characteristics to the underlying connections. Specifically, stochastic system theory which characterizes the active variable of indeterminate probabilistic systems (in terms of a probability density function, correlations, entropy, and the like) were applied to the neurons in order to elucidate the inferences concerning interconnections and emergent properties of neural networks on the basis of activity correlations (*cooperative processes*) among the constituting units.

A proposal to describe the global electrical activity of the neural complex in terms of theoretical concepts similar to those used to describe the global properties of materials in statistical mechanics was suggested in 1948 by Norbert Weiner in his classical book *Cybernetics* [9] — the study of self-regulating systems. The underlying basis for his proposed analogy is founded on the following considerations: The theoretical science of *statistical mechanics* makes inferences concerning global properties and constraints based on the aggregate of the physical rules describing the individual molecular interactions. The vast number of neurons which interact with each other represent analogously the interacting molecules, and hence the pertinent similarity permits an inferential strategy on the global properties of the neurons, as is possible with the molecular system. "Intuitively Wiener must have realized that statistical mechanics is ideally suited to analyze collective phenomena in networks consisting of very many relatively simple constituents."

Once the aforesaid analogy was recognized, the various forms of theorizing the interactions (or the so-called cooperative processes), namely, the *Lagrangian*, the *Hamiltonian*, the *total energy*, the *action*, and the *entropy* as defined in classical mechanics and in the thermodynamics of solids and fluids also were extended to the neural activity. Hence, the essential consideration that the systems tend towards the extrema of the aforesaid functions (for example, minimization of total energy or maximization of entropy), as adopted commonly in statistical mechanics, indicates the application of similar approaches to the neural system. This also permits the evaluation of the global behavior of "rich and complicated" systems (such as the neural assembly) by a single global function advocated by a single principle.

The organization of neurons is a collective enterprise in which the neural activity refers to a cooperative venture involving each neuron interacting with many of its neighbors with the culmination of such activities in a (*dichotomous*) *all-or-none* response to the incoming stimuli. In this cooperative endeavor, the interaction between the neurons is mediated by the (*all-or-nothing*) impulses crossing the synapses of a closely linked cellular anatomy randomly, constituting a *collective movement* of a stochastical activity.

Essentially, in the neuronal conduction process as discussed in the earlier chapters, a cellular neuron (in a large assembly of similar units) is activated by the flow of chemicals across synaptic junctions from the axon leading from other neurons; and the resulting output response can be viewed as either *excitatory* or *inhibitory* postsynaptic potential. If the gathered potentials from all the incoming synapses exceed a threshold value, the neuron *fires* and this excitatory process sets an action potential to propagate down one of the output axons (which eventually communicates with other neurons, *via* synaptic tributory branches). After firing, the neuron returns to its quiescent (*resting*) potential and is sustained in that condition over a *refractory period* (of about several milliseconds) before it can be excited again. The firing pattern of the neurons is governed by the topology of interconnected neurons and the collective behavior of the neuronal activity.

The threshold based input-output response of a neuron was represented as a simple, two-state logical system — *active* and *inactive* — by McCulloch and Pitts in 1943 [7]. As indicated in the previous chapters, this is known as the *formal* or *logical* or *mathematical* neuron model. Denoting the state of a neuron (at time t) by a variable S_i which can have two values: $S_i = +1$ if it is active and $S_i = -1$ if it is inactive, and referring to the *strength of synaptic connection* between two arbitrary cells i and j as W_{ij}, the sum total of the stimuli at the ith neuron from all the others is given by $\Sigma_j W_{ij} S_j$. This is the postsynaptic potential, or in physical terms a local field h_i. Setting the threshold for the ith neuron as $T_H{}^i$, then $S_i = +1$ if $\Sigma W_{ij} S_j > T_H{}^i$ and $S_i = -1$ if $\Sigma W_{ij} S_j < T_H{}^i$. Together with an external bias θ_i added to the summed up stimuli at the ith neuron, the neuronal activity can be specified by a single relation namely, $S_i (\Sigma W_{ij} S_j - \theta_i) > 0$; and a corresponding Hamiltonian for the neural complex can be written as:

$$H_N = - (\Sigma\, W_{ij} S_i S_j) - (\Sigma\, \theta_i S_i) \qquad (5.1)$$

with $\{S_i\}$ denoting the state of the whole system at a given instant of time; and both S_i and S_j are the elements of the same set $\{S_i\}$.

73

The aforesaid mathematical depiction of simple neural activity thus presents a close analogy between the neural network and *magnetic spin model(s)*. That is, the neurons can be regarded as analogs of the *Ising spins* (see Appendix A) and the strengths of the synaptic connections are analogous to the strengths of the *exchange interactions* in spin systems. The concept of interaction physics as applied to a system of atomic magnetic dipoles or spins refers to the phenomenon namely, atoms interact with each other by inducing a magnetic field at the location of other (neighboring) atoms which interact with its spin. The total local magnetic field at the location of an atom i is equal to $\Sigma_{ij}W_{ij}S_j$ where W_{ij} is the dipole force, and diagonal term $j = i$ (self-energy) is not included in the sum. Further, Newton's third law, namely, *action* equates to *reaction*, ensures the coupling strengths W_{ij} being symmetric. That is, $W_{ij} = W_{ji}$. If all W_{ij} are positive, the material is ferromagnetic; if there is a regular change of sign between neighboring atoms, it refers to antiferromagnetism. If the signs and absolute values of the W_{ij} are distributed randomly, the material is called *spin-glass*. The ferromagnetic case corresponds to a neural network that has stored a single pattern. The network which has been loaded with a large number of randomly composed patterns resembles a spin glass.

Synaptic activity in neuronal systems being *excitatory* or *inhibitory*, the competition between these two types of interactions can be considered as similar to the competition between the *ferromagnetic* and *antiferromagnetic* exchange interactions in *spin-glass* systems. That is, the dichotomous "all-or-none" variables of neurons correspond to $S_i = \pm 1$ Ising spins where i labels the neurons, and range between 1 and N determines the size of the network. Further, the threshold condition stipulated for the neural complex can be regarded as the analog of the condition for *metastability* against single-spin flips in the Ising model (except that in a neural complex the symmetry relation, namely, $W_{ij} = W_{ji}$, does not necessarily hold). The evolution of the analogical considerations between interconnected neurons and the magnetic spins is discussed in detail in the following sections.

5.2 Cragg and Temperley Model

In view of the above analogical considerations between neurons and magnetic spins, it appears that the feasibility of applying quantum theory mathematics to neurobiology was implicitly portrayed by Gabor [10] as early as in 1946 even before Weiner's [9] suggestion on cybernetic aspects of biological neurons. As indicated by Licklider [75], "the analogy ... [to] the position-momentum and energy-time problems that led Heisenberg in 1927 to state his uncertainty principle ... has led Gabor to suggest that one may find the solution [to the problems of sensory processing] in quantum mechanics."

In 1954, Cragg and Temperley [32] were perhaps the first to elaborate and examine qualitatively the possible analogy between the *organization of neurons* and the kind of interaction among atoms which leads to the cooperative processes in physics. That is, the purported analogy stems from the fact that large assemblies of atoms which interact with each other correspond to the collective neural assembly exhibiting cooperative activities through interconnections.

As explained before, in the case of an assembly of atoms, there is an explicit degree of interaction manifesting as the phenomenon of ferromagnetism; and such an interaction between atomic magnets keeps them lined up (polarized) below a specific temperature (known as the *Curie point*). (Above this temperature, the increase in thermal agitation would, however, throw the atomic magnets out of alignment; or the material would abruptly cease to be ferromagnetic).

The membrane potential of each neuron due to the interconnected configuration with the other cells could likewise be altered as a result of changes in membrane potential of any or all of the neighboring neurons. The closely packed neurons, as observed by Cragg and Temperley, hence permit the application of the theory of cooperative processes in which the cells are at two states of energy levels; and the interaction between the cells introduces a correlation between the states occupied by them. The whole assembly has a fixed amount of energy which is conserved but with no restriction to change from one configuration to the other.

The interaction across the whole assembly also permits a proliferation of state changes through all the possible neuronal configurations. Each configuration, however, has a probability of occurrence; and hence the average properties of the whole assembly refer to the weighted averaging over all the possible configurations.

In the mediating process by all-or-nothing impulses as mentioned before, the pertinent synaptic interaction could be either excitatory (*hypopolarized*) or inhibitory (*hyperpolarized*) interaction perceived as two different directional (ionic) current flows across the cellular membrane. Suppose the excitatory and inhibitory interactions are exactly balanced. Then the overall effect is a null interaction. If, on the other hand, the inhibitory process dominates, the situation is analogous to antiferromagnetism which arises whenever the atomic interaction tends to set the neighboring atomic magnets in opposite directions. Then, on a macroscopic scale no detectable *spontaneous magnetism* would prevail. In a neurological analogy, this corresponds to a zero potential difference across the cellular domains.

It has been observed in neurophysiological studies that the hyperpolarization (inhibitory) process is relatively less prominent than the excitatory process; and the collective neuronal process would occur even due to an asymmetry of the order 1002:1000 in favor of the excitatory interactions.

75

Cragg and Temperley hypothesized that a (large) set of M neurons analogously corresponds to a set of M atoms, each having spins ±1/2. A neuron is regarded as having two (dichotomous) states distinguished by the presence (all) or absence (none) of an action potential which may be correlated with the two independent states possible for an atom in a state having spin 1/2 with no further *degeneracy*[*] due to any other factors.

5.3 Concerns of Griffith

Almost a decade later, the qualitative analogy presented by Cragg and Temperley received a sharp criticism from Griffith [13,14] who observed that Cragg and Temperley had not defined the relation to ferromagnetic material in sufficient detail for one to know whether the analogies to macroscopic magnetic behavior should actually hold in the neural system. In the pertinent study, the neural assembly representing an aggregate of M cells with the dichotomous state of activity has 2^M possible different states which could be identified by $S = 1, ..., 2^M$ associated with an M-dimensional hypercube; and Griffith observed that a superficial analogy here with the quantum statistical mechanics situation corresponds to a set of M subsystems, each having two possible quantum states as, for example, a set of M atoms each having a spin 1/2 (with no additional degeneracy due to any other possible contributing factors).

Each of the 2^M states has a definite successor in time so that the progress of the state-transitional process (or the neuronal wave motion) can be considered as a sequence $i_2 = z(i_1) \rightarrow i_3 = z(i_1) \rightarrow$ ... and so on.

Regarding this sequence, Griffith pointed out that in the terminal cycle of the state-transitional process, there are three possible situations, namely, a state of equilibrium with a probability distribution $\rho(S_0)$; and the other two are dichotomous states, identified as $S_1 \Rightarrow + S_U$ and $S_U \Rightarrow - S_L$ with the statistics depicted by the probability distributions $\rho(S_1)$ and $\rho(S_2)$, respectively.

In computing the number of states which end up close to the equilibrium at the terminal cycle, Griffith indicated the following fundamental difference between the neural and the quantum situation: From the quantum mechanics point of view, the state-transitional probabilities (due to an underlying barrier potential ϕ) between two states S_1 and S_2

---

[*] Here *degeneracy* refers to several states having the same energy levels. It is quantified as follows: A microcanonical system (in which all energies in the sum of the states are equal) has a *partition function* $Z = \Sigma_v \exp(-\beta E_v) = M_F \exp(-\beta U)$ where M_F is called the *degeneracy* or spectral multiplicity of the energy level and U is the *average energy* of the system. The partition function which essentially controls the average energy by the relation $U = -\partial(\ln F)/\partial\beta$ can be written as $Z = M_F \exp(-\beta U)$.

with probabilities $\rho_{1,2}$ (and corresponding wave functions $\Psi_{1,2}$) specified with reference to the equilibrium state, namely, S_0 (with $\rho_0 \Rightarrow \Psi_0$) are equal in both directions. This is because they are proportional, respectively, to the two sides of the equation given by :

$$|<\Psi_0 \,|\phi|\, \Psi_{12}>|^2 \;=\; |<\Psi_{12} \,|\phi|\, \Psi_0>|^2 \qquad (5.2)$$

The above relation is valid inasmuch as ϕ is Hermitian.[*] In the case of neural dynamics, however, Griffith observed that the possibilities of $i_2 = z(i_1)$ and $i_1 = z(i_2)$ are rather remote. That is, there would be no *microscopic reversibility*. There could only be a natural tendency for the microscopic parameter $\rho_{1,2}$ to move near to ρ_0 and "there would not seem to be any very obvious reason for the successor function z to show any particular symmetry."

In essence, Griffith's objection to symmetry in the synaptic weight space has stemmed from his nonconcurrence with the theory proposed by Cragg and Temperley to consider the neural networks as aggregates of interacting spins (as in ferromagnetic materials). In enunciating a correspondence between the neural networks and magnetic spin systems, as done by Cragg and Temperley, it was Griffith's opinion that the Hamiltonian of the neural assembly "is totally unlike a ferromagnetic Hamiltonian ... the (neural) Hamiltonian has the undesirable features of being intractably complicated and also non-hermitian. ... [hence] the original analogy [between neural network and magnetic spin system] is invalid. ... This appears to reduce considerably the practical value of any such analogy."

Notwithstanding the fact that the spin-glass analogy extended to neuronal activity was regarded by Griffith as having no "practical value", a number of studies have emerged in the last two decades either to justify the analogy or to use the relevant parallelism between the spin-glass theory and neural dynamics in artificial neural networks. Such contributions have stemmed from cohesive considerations related to statistical physics, neurobiology, cognitive and computer sciences, and relevant topics which cover the general aspects of time-dependent problems, coding and retrieval considerations, hierarchical organization of neural systems, biological motivations in modeling neural networks analogous to spin glasses, and other related problems have been developed. The analogy of the neural complex with spin systems had become an important topic of interest due to the advances made in understanding the thermodynamic properties of

---

[*] If A is an $m \times n$ matrix, $A = [a_{ij}]_{(mn)}$, A is Hermitian if $A = A^*$, where $A^* = [b_{ij}]_{(nm)}$, with $b_{ij} = \bar{a}$ (the notation $\bar{a}$ denotes the complex conjugate of the number a)

disordered systems of spins, the so-called *spin glasses* over the last scores of years. When the pertinent results are applied to neural networks, the *deterministic evolution law* of updating the network output is replaced by a *stochastic* law where the state variable of the cell (at a new instant of time) is assigned according to a probabilistic function depending on the intensity of the synaptic input. This probabilistic function is dictated by the pseudo-temperature concepts outlined in Chapter 4. The stochastical evolution law pertains to the features of real neurons wherein spontaneous firing without external excitation may be encountered leading to a persistent noise level in the network.

Among the existing studies, more basic considerations into the one-to-one analogy of spin-glass theory and neuronal activity were considered exclusively in detail by Little [33] and by a number of others, a chronological summary of which is presented in the following sections.

5.4 Little's Model

Subsequent to Griffith's verdict on the spin-glass model of the neural complex, Little in 1974 [33] demonstrated the existence of *persistent states* of firing patterns in a neural network when a certain transfer matrix has approximately *degenerate* maximum eigenvalues.* He demonstrated a direct analogy of the persistence in neuronal firing patterns (considered in discrete time-steps) to the *long-range spatial order* in an Ising spin crystal system; and the order-disorder situations in the crystal lattice are dictated by the thermodynamic considerations specified by the system temperatue T. The ordered phase of the spin system occurs below a critical temperature (T_C), well known as the *Curie point*. Analogously, a factor (β) representing the temperature of the neural network is assumed in Little's model for the transfer matrix that depicts the persistent states. The approach envisaged by Little can be summarized as follows.

In proposing the analogy between a network of neurons and the statistical mechanics-based Ising spin system, Little considered that the temporal development of the network (in discrete time-steps) corresponds to a progression across one dimension in the Ising lattice. In the Ising model as indicated earlier, the spin S_i at each lattice site i can take only two different orientations, up and down, denoted by $S_i = +1$ (up) and $S_i = -1$ (down). The analogy to a neural network is realized by identifying each spin with a neuron and associating the upward orientation $S_i = +1$ with the active state 1 and the downward orientation $S_i = -1$ with the resting state 0.

---

* A matrix which has no two eigenvalues equal and which has, therefore, just as many distinct eigenvalues as its dimension is said to be *nondegenerate*. That is, if more than one eigenvector has the same eigenvalue, the matrix is *degenerate*.

Further, he suggested that certain networks of neurons could undergo a transition from a *disordered state* to an *ordered state* analogous to the Ising-lattice phase transition. Since this implies temporal correlations, he pointed out that these ordered states might be associated with *memory*, as well.[*]

Little's model is a slightly more complex description of the logical or formal neuron due to McCulloch and Pitts [7]. It accounts for the chemical transmission at synapses. Its model parameters are the synaptic potentials, the dichotomous thresholds, and a quantity β which represents the net effect on neural firing behavior of variability in synaptic transmission. In Little's model, the probability of firing ranges from 0 to 1 and is a function of the difference between the total membrane potential and the threshold. Further, the probability of each neuronal firing is such that the time evolution of a network of Little's neurons is regarded as a *Markov process*.

To elaborate his model, Little considered an isolated neural network. He analyzed the state of the system pictured in terms of the neurons being active or silent at a given time and looked at the evolution of such a state in discrete time-steps τ greater than the refractory period (τ_R) for long periods (greater than 100 τ), searching for correlations of such states. He showed that *long-time* correlation of the states would occur if a certain transfer matrix has approximately degenerate maximum eigenvalues. He suggested that these persistent states are associated with a short-term memory.

Little related the (three-dimensional) isolated system of M neurons quantized in discrete time-steps to the spin states of a two-dimensional Ising model of M spins with no connections between spins in the same row. The state of the neural network at an instant of time corresponds to the configuration of spins in a row of the lattice of the crystal. The state of the neurons after one time step τ corresponds to the spin configuration in the next row.

---

[*] The *memory*, in general, can be classified on the basis of three time scales, namely:
Short-term memory: This is an image-like memory lasting from a fraction of a second to seconds. In reference to the neuronal assembly, it depicts to a specific firing pattern or a cyclic group of patterns persisting in the active states over this time period after an excitation by a strong stimulus (which will override any internal firing trend in the network, forcing it to have a specific pattern or a set of patterns. Such a stimulus would facilitate enhancement of certain neurocortical parameters associated with firing).
Intermediate memory: This could last up to hours during which time imprinting into long-term memory can be affected by drugs or electric shock, etc. The synaptic parameters facilitated still may cause the reexcitation of a pattern in the network.
Long-term memory: This refers to an almost permanent memory depicting plastic or permanent changes in synaptic strength and/or growth of new synapses, but still the facilitated parameters may enable pattern reexcitation.

The potential ϕ_{ij} accruing at time t to the ith neuron from its jth synapse due to the firing of the jth neuron at time $(t - \tau)$ was related by analogy to the energy of spin interactions ϕ_{ij} between the ith and j^{th} spins on adjacent rows only. (Unlike the Ising problem, the neural connections, however, are not symmetric.) It was assumed that probability of neuronal firing was given by an expression similar to the *partition function* of the spin system. The persistent (in time) states of the neural network were therefore studied in the usual approach of analyzing long-range (spatial) order in the spin systems.

A suitable product of the probabilities for firing or not firing will constitute the transition matrix elements for the neuronal state configurations at successive time intervals. As is well known in Ising model calculations, degeneracy of the maximum eigenvalues of the transition matrix is associated with condensation of the spin system below the Curie point temperature and corresponds to a new phase and long-range order. Hence, a factor (β) representing the (pseudo) temperature of the neural network appears inevitably in the transition matrix of Little's model.

Considering an *isolated* network or ganglion of M neurons, the potential of a neuron is determined effectively by the integrated effects of all excitatory postsynaptic potentials as well as inhibitory postsynaptic potentials received during a period of summation (which is in the order of a few milliseconds). The neurons are assumed to fire at intervals of τ which is the order of the refractory period τ_R, also a few milliseconds in duration. Conduction times between neurons are taken to be small. At each time interval, the neurons are assumed to start with a clear slate (each neuron's potential is reset to its resting value). This corresponds to a first-order *markovian process*. All properties of the synaptic junctions are assumed to be constant over the time scales of interest (implying an *adiabatic* hypothesis).

Using the terminology of a quantum mechanical spin system, the state of the brain (as determined by the set of neurons that have fired most recently and those that have not) is the configuration represented by ψ at the (discrete) time t .

$$\psi = |S_1, S_2, ..., S_M> \qquad (5.3)$$

$S_i = +S_U(= +1)$, if the ith neuron fires at time t or $S_i = -S_L (= -1)$ and if it is silent corresponding to the *up* and *down* states of a spin system.[*] Let ϕ_{ij} be the postsynaptic potential of the ith neuron due to the firing of

---

[*] For a particular configuration of spins, say $\{S_1, S_2, ... \}$, $S_i = +S_U$ (or +1) refers to the spin being *up;* and, when $S_i = -S_L$ (or –1), the spin is labeled as *down.*

the jth neuron. Thus the total potential of the ith neuron is given by $\Sigma^M_{j=1} \phi_{ij} (S_j + 1)/2$. If the total potential exceeds a threshold value ϕ_{pBi} (the barrier potential, possibly independent of i), the neuron will probably fire; and the probability of firing is assumed by Little, in analogy with the spin system, to be:

$$\rho_i(+S_U) = [\exp\{-\beta[\sum^M_{j=1}[\phi_{ij}(S_j + 1)/2] - \phi_{pBi}] + 1\}]^{-1}$$

$$(5.4)$$

for time $t' = (t + \tau)$.

The *temperature* factor $\beta = 1/k_BT$ in the spin system (with k_B denoting the [pseudo] Boltzmann constant) is related to the uncertainty in the firing of the neuron. The probability of not firing is $1 - \rho_i(+S_U)$. Thus, the probability of obtaining at time $(t + \tau)$ the state $\psi' = |S_1'S_2' ... S_M'>$, given the state $\psi = |S_1,S_2, ..., S_M>$ at one unit of time τ preceding it, is given by:

$$\psi' T_M \psi = \prod^M_{i=1} [\exp\{-\beta S_i' (\sum^M_{j=1} \phi_{ij}[(S_j + 1)/2] - \phi_{pBi}) + 1\}]^{-1}$$

$$= \prod^M_{i=1} \exp\{\beta S_i'/2\}[\Phi(S_j)]/\prod^M_{i=1} \sum_{S'=\pm1} \exp\{\beta S_i'/2\}[\Phi(S_j)]$$

$$(5.5)$$

where $\Phi(S_j) = \Sigma^M_{j=1} [\phi_{ij}(S_j + 1)/2] - \phi_{pBi}$.

It may be noted that this expression is very similar to that occurring in the study of propagation of order in crystals with rows of atomic spins. Further, the interaction is between spins with dashed indices and undashed indices, that is, for spins in adjacent rows and not in the same row (or different time steps in the neural network). Long-range order exists in the crystal whenever there is a correlation between distant rows. Ferromagnetism sets in when long range order gets established below the Curie temperature of the spin system.[*] There are 2^M possible spin states as

_____

[*] A crystal undergoes a *phase transition* from the paramagnetic state to the ferromagnetic state at a sharply defined temperature (Curie point). At this transition temperature, the properties of the matrix change so that the ferromagnetic state information contained in the first row of the crystal propagates throughout the crystal. In a neural network, this represents analogously the capability of the network to sustain a persistent firing pattern.

specified by Equation (5.3); and likewise 2^M x 2^M matrix elements as specified by Equation (5.5) which constitute a 2^M x 2^M transfer matrix T_M. This long range order is associated with the degeneracy of the maximum eigenvalue of T_M. In the neural problem, the firing pattern of the neuron at time t corresponds to the up-down state of spin in a row and the time steps are the different rows of the crystal problem. Since the neuronal ϕ_{ij} is not equal to ϕ_{ji}, the matrix T_M from Equation (5.5) is not symmetric as in the spin problem and thus cannot always be diagonalized. This problem can, however, be handled in terms of principal vectors rather than eigenvector expansions in the spin system without any loss of generality.

Let $\Psi(\alpha_i)$ represent the 2^M possible states Equation (5.3). Then the probability of obtaining state $\Psi(\alpha')$ having started with $\Psi(\alpha)$ (m time intervals [τ] earlier) can be written in terms of the transfer matrix of Equation (5.5) as:

$$\Psi(\alpha') \, T_M^m \, \Psi(\alpha) \tag{5.6}$$

As is familiar in quantum mechanics, $\Psi(\alpha)$ can be expressed in terms of the 2^M (orthonormal) eigenvectors ϑ_r (with eigenvalues λ_r) of the operator T_M. Each ϑ_r has 2^M components, one for each configuration α:

$$\Psi(\alpha) = \sum_{r=1}^{2^M} \vartheta_r(\alpha) \tag{5.7}$$

Hence

$$<\Psi(\alpha') \, |T_M| \, \Psi(\alpha)> = \sum_r \lambda_r \, \vartheta_r(\alpha') \, \vartheta_r(\alpha) \tag{5.8}$$

Little's approach is concerned with the probability $\Gamma(\alpha_1)$ of finding a particular state α_1 (after m time-steps) starting with an arbitrary initial state. Analogous to the method of using *cyclic boundary conditions* in the spin problem (in order to simplify derivations, yet with no loss of generality), it is assumed that the neural system returns to the initial conditions after M_0 (>>m) steps. Hence, it follows that:

$$\Gamma(\alpha_1) = \sum_r \lambda_r^{M_0} \, \vartheta_r^2(\alpha) / \sum_r \lambda_r^{M_0} \tag{5.9}$$

(which is independent of m). Now, with the condition that the maximum eigenvalue λ_{max} is nondegenerate, Equation (5.9) reduces to:

$$\Gamma(\alpha_1) = \vartheta_{max}^2(\alpha_1) \qquad\qquad (5.10)$$

Hence, it follows that the probability of obtaining the state α_2 after ℓ time steps, given α_1 after m steps, is given by:

$$\Gamma(\alpha_1, \alpha_2) = \vartheta_{max}^2(\alpha_1)\, \vartheta_{max}^2(\alpha_2) = \Gamma(\alpha_1)\,\Gamma(\alpha_2) \qquad (5.11)$$

indicating no correlation. However, if the maximum eigenvalue λ_{max} is degenerate, the factorization of $\Gamma(\alpha_1, \alpha_2)$ is not possible; and there will be correlation in time in the neuronal-firing behavior. This type of degeneracy occurs in the spin system for some regions of a $\beta - (\phi_{ij} - \phi_{pBi})$ plane and refers to the transition from the paramagnetic to the ferromagnetic phase. Relevant to the neural complex, Little suggests that such time ordering is related to short-term memory. Since time correlations of the order less than or equal to a second are of interest in the neural dynamics, a practical degeneracy will result if the two largest λ's are degenerate to ~1%.

In the above treatment, the parameter β assumed is arbitrary. However, this β could represent all the *spread* in the uncertainty of the firing of the neuron. This has been demonstrated by Shaw and Vasudevan [76] who suggested that the *ad hoc* parameter β in reality relates to the fluctuations governing the total (summed up) potential gathered by the neuron in a time-step (which eventually decides the state of the neuron at the end of the time-step as well). The relevant analysis was based on the probabilistic aspects of synaptic transmission, and the temperature-like factor or the pseudo-temperature universe β (= $1/k_B T$ in the Ising model) was termed as a *smearing parameter*.

Explicitly, this smearing parameter (β) has been shown equal to $2\sqrt{2/\pi}/\Delta$, where Δ is a factor decided by the *gaussian statistics* of the action potentials and the *poissonian process* governing the occurrence rate of the quanta of chemical transmitter (ACh) reaching the postsynaptic membrane (and hence causing the postsynaptic potential). The relevant statistics indicated refer to the variations in size and the probability of release of these quanta manifesting (and experimentally observed) as fluctuations in the postsynaptic potentials.

In a continued study on the statistical mechanics aspects of neural activity, Little and Shaw [77] developed a model of a large neural complex (such as the brain) to depict the nature of short- and long-term memory. They presumed that memory results from a form of synaptic strength modification which is dependent on the correlation of pre- and postsynaptic neuronal firing; and deduced that a reliable, well-defined behavior of the assembly would prevail despite of noisy (and hence random) characteristics of the membrane potentials due to the fluctuations (in the number and size) of neurochemical transmitter molecules (ACh quanta) released at the synapses. The underlying basis for their inference is that the neuronal collection represents an extensive assembly comprised of a (statistically) large number of cells with complex synaptic interconnections, permitting a stochastically viable proliferation of state changes through all the possible neuronal configurations (or patterns of neural conduction).

In the relevant study, the pertinent assumption on modifiable synapses refers to the *Hebbian learning process*. In neurophysiological terms it is explicitly postulated as: "When an axon of cell A is near enough to excite a cell B and repeatedly or persistently takes part in firing it, some growth process or metabolic change takes place in one or both cells such that A's efficiency as one of the cells firing B is enhanced. [19]"

In further studies concerning the analogy of neural activity *versus* Ising spin model, Little and Shaw [78] developed an analytical model to elucidate the memory storage capacity of a neural network. They showed thereof that the memory capacity is decided by the (large) number of synapses rather than by the (much smaller) number of neurons themselves; and by virtue of this large memory capacity, there is a storage of information generated *via* patterns of state-transition proliferation across the neural assembly which evolves with time. That is, considering the long-term memory model, the synaptic strengths cannot be assumed as time-invariant. With the result, a modified Hebb's hypothesis, namely, that the synaptic changes do occur as a result of correlated pre- and postneuronal firing behavior of the linear combinations of the (spatial) firing pattern, was suggested in [78]. Thus, the relevant study portrayed the existence of possible spatial correlation (that is, firing correlation of neighboring neurons, as evinced in experimental studies) in a neural assembly. Also, such correlations resulting from the linear combination of firing patterns corresponds to M^2 transitions, where M is the number of neurons; and with every neuron connected to every other neuron, there are M^2 number of synapses wherein the transitions would take place.

The aforesaid results and conclusions of Little and Shaw were again based mainly on the Ising spin analogy with the neural system. However, the extent of their study on the linear combination firing patterns from the statistical mechanics point of view is a more rigorous, statistically involved task warranting an analogy with the three-dimensional Ising problem unfortunately, this remains unsolved to date. Nevertheless, the results of Little and Shaw based on the elementary Ising spin model, indicate the possibility of spatial firing correlations of neighboring neurons which have been confirmed henceforth *via* experiments using two or more closely spaced microelectrodes.

5.5 Thompson and Gibson Model

Thompson and Gibson in 1981 [37] advocated in favor of Little's model governing the probabilistic aspects of neuronal-firing behavior with the exception that the concept of long-range order introduced by Little is considered rather inappropriate for the neural network; and they suggested alternatively a more *general definition of the order*. The relevant synopsis of the studies due to Thompson and Gibson follows.

Considering a spin system, if fixing the spin at one lattice site causes spins at sites far away from it to show a preference for one orientation, it refers to the long-range order of the spin system. To extend this concept to the neural assembly, it is necessary first to consider the Ising model of the two-dimensional ferromagnet in detail. In the Ising spin system, a regular lattice of spins $S_i = \pm 1$ with an isotropic nearest-neighbor interaction is built by successive addition of rows, each consisting of M spins, where M is finite. The probability distribution of spins in the $(m + 1)^{th}$ row depends only on the distribution in the mth row, depicting a Markov process with a transition matrix T_M. In this respect, the neural network and spin structure are formally analogous; and the time-steps for the neural network correspond to the spatial dimension of the spin lattice, as discussed earlier.

In the spin problem, the transition matrix T_M is strictly a positive stochastic matrix for all positive values of the temperature T such that the long-range order for any finite spin system with $T > 0$ is not feasible. However, in the limit as $M \to \infty$, the largest eigenvalue of T_M is asymptotically degenerate provided $T < T_C$, (T_C being the Curie point). In this case, T_M^m no longer approaches a matrix with equal components when m becomes arbitrarily large. This infinite two-dimensional spin system undergoes a sharp phase transition at T_C. For $T > T_C$, there is no long range order and each spin is equally likely to be up or down, whereas for $T < T_C$, there is a long-range order and the spins are not randomly oriented (see Appendix A).

The nearest-neighbor spin-spin interactions in a ferromagnetic system are symmetric as discussed earlier, and the effect that one spin has on the orientation of any other spin depends only on their spatial separation in the lattice. Hence, the successive rows of the spin system can be added in any direction; however, considering the neural system, the analogous time development steps have only a specific forward direction. *That is, the neuronal interaction is inherently anisotropic.* The state of the neuron at any time is determined by the state of all the neurons at the previous time. This interaction for a given neuron can be distinctly different and unique when considered with other neurons. It also depends on the synaptic connectivity of the particular network in question. Generally, the interaction of the jth neuron with the ith neuron is not the same as that of the ith neuron with the jth ; and *the transition matrix* T_M *is, therefore, nonsymmetric.* That is, the synaptic connections are generally not symmetric and are often maximally asymmetrical or unidirectional. (Müller and Reinhardt [1] refers to such networks as *cybernetic networks* and indicate the feed-forward, layered neural networks as the best-studied class of cybernetic networks. They provide an optimal reaction or answer to an external stimulus as dictated by a supervising element [such as the brain] in the self-control endeavors. Further, as a result of being asymmetric the theories of thermodynamic equilibrium systems have no direct bearing on such cybernetic networks.)

Thompson and Gibson hence declared that the spin system definition of long-range order is rather inapplicable to neural assembly due to the following reasons: (1) Inasmuch as the interaction between different neurons have different forms, any single neuron would not influence the state of any other single neuron (including itself) at a later time; and (2) because the transition matrix is asymmetric (not necessarily diagonalizable) in a neuronal system, the long-range order does not necessarily imply a tendency for the system to be in a particular or persistent state. (On the contrary, in a spin system, the order is strictly a measure of the tendency of the spins aligned in one direction in preference to random orientation.)

As a result of the inapplicability of the spin-system based definition of *long-range order* to a neural system, Thompson and Gibson proposed an alternative definition of long-range order which is applicable to both the spin system as well as the neuronal system. Their definition refers to the *order* of the system applied to a *moderate time scale* and not for the *long-range epoch.* In this moderate time-frame order, plastic changes in synaptic parameters would be absent; and by considering the neural network as a finite (and not as an arbitrarily large or infinite system), the *phase transition process* (akin to that of the spin system) from a disordered to an ordered state would take place in a continuous graded fashion rather than as a sharp transition. Thus, the spin system analogy is still applicable to the neural system provided a finite system assumption and moderate time-scale order are attributed explicitly to the neuronal state transition process.

Thompson and Gibson further observed that the aforesaid gradual transition refers to the factor β being finite-valued. If $\beta \to \infty$, it corresponds to the McCulloch-Pitts regime of the neuron being classified as a logical or formal neuron. It also implicitly specifies the network of Little having a long-range order.

In the continuous/graded state transition corresponding to a moderate time-scale order, the firing pattern could be of two types: (1) The *burst discharge pattern* characterized by the output of an individual neuron being a series of separated bursts of activity rather than single spikes that is, the network fires a fixed pattern for some time and then suddenly changes to a different pattern which is also maintained for many time steps and (2) *the quasi-reverberation* pattern which corresponds to each neuron making a deterministic fire or no-fire decision at multiples of a basic unit of time; and a group of such neurons may form a closed, self-exciting loop yielding a cyclically repeating pattern called *reverberation*. Thompson and Gibson identified the possibility of the existence of both patterns as governed by the markovian statistics of neuronal state transition. Their further investigations on this topic [79], with relevance to Little's model, has revealed a single model neuron can produce a wide range of average output patterns including *spontaneous bursting* and *tonic firing*. Their study was also extended to two neuron activities. On the basis of their results, they conclude that Little's model "produces a remarkably wide range of physically interesting average output patterns ... . In Little's model, the most probable behavior [of the neuronal network] is a simple consequence of the synaptic connectivity ... That is, the type of each neuron and the synaptic connections are the primary properties. They determine the most likely behavior of the network. The actual output could be slightly modified or stabilized as a result of the various secondary effects" [such as accommodation or postinhibitory rebound, etc.].

5.6 Hopfield's Model

As observed by Little, the collective properties of a large number of interacting neurons compare to a large extent with the "physical systems made from a large number of simple elements, interactions among large numbers of elementary components yielding collective phenomena such as the stable magnetic orientation and domains in a magnetic system or the vortex patterns in a fluid flow." Hence, Hopfield in 1982 [31] asked a consistent question, "Do analogous collective phenomena in a system of simple interacting neurons have useful *computational* correlates?" Also, he examined a new modeling of this old and fundamental question and showed that "important computational properties" do arise.

The thesis of Hopfield compares neural networks and physical systems in respect to emergent collective computational abilities. It follows the time evolution of a physical system described by a set of general coordinates, with a point in the state-space representing the instantaneous condition of the system; and, this state-space may be either continuous or discrete (as in the case of M Ising spins depicted by Little).

The input-output relationship for a neuron prescribed by Hopfield on the basis of collective properties of neural assembly has relevance to the earlier works due to Little and others which can be stated as follows [31]: "Little, Shaw, and Roney have developed ideas on the collective functioning of neural nets based on "on/off" neurons and synchronous processing. However, in their model the relative timing of action potential spikes was central and resulted in reverberating action potential trains. Hopfield's model and theirs have limited formal similarity, although there may be connections at a deeper level."

Further, considering Hopfield's model, when the synaptic weight (strength of the connection) W_{ij} is symmetric, the state changes will continue until a local minimum is reached. Hopfield took random patterns where $\xi_i^\mu = \pm 1$ with probability 1/2 and assumed $W_{ij} = [\Sigma_\mu \, \xi_i^\mu \xi_j^\mu]/N$, $(i, j) \in N$ and allowed a sequential dynamics of the form $S_i(t + \Delta t) =$ Sgn $[h_i(t)]$ where Sgn(x) is the sign of x and $h_i = \Sigma_j \, W_{ij} S_j$ and represents the postsynaptical or the local field. Hopfield's dynamic is equivalent to the rate that the state of a neuron is changed, or a spin is flipped, iff the energy $H_N = -\Sigma_{i \neq j} \, W_{ij} S_i S_j$ is lowered. That is, the Hamiltonian H_N is the so-called Lyapunov function for the Hopfield dynamics which converges to a local minimum or the ground state. Or, the equations of motions for a network with symmetric connections $(W_{ij} = W_{ji})$ always lead to a convergence to stable states in which the outputs of all neurons remain constant. Thus the presumed symmetry of the network is rather essential to the relevant mathematics. However, the feasibility of the existence of such a symmetry in real neurons has rather been viewed with skepticism as discussed earlier.

Hopfield has also noted that real neurons need not make synapses both of i → j and j → i, and questioned whether $W_{ij} = W_{ji}$ is important *vis-a-vis* neuronal activity. He carried out simulations with only one ij connection, namely, $W_{ij} \neq 0$, $W_{ji} = 0$, and found that without symmetry the probability of making errors increased though the algorithm continued to generate stable minima; and there was a possibility that a minimum would be only metastable and be replaced in time eventually by another minimum. The symmetric synaptic coupling of Hopfield, however provoked a great deal of criticism as being biologically unacceptable; as Toulouse [80] points out Hopfield's strategy was a "very clever step backwards."

In a later work, Hopfield [36] introduced electronic circuit modeling of a larger network of neurons with graded response (or sigmoidal input-output relation) depicting *content-addressable memory* based on the collective computational properties of two-state neurons. The relevant model facilitates the inclusion of propagation delays, jitter, and noise as observed in real neurons. The corresponding stochastic algorithm is asynchronous as the interaction of each neuron is a stochastic process taking place at a mean rate for each neuron. Hence, Hopfield's model, in general, differs from the synchronous system of Little which might have additional collective properties.

Pursuant to the above studies on neural activity *versus* statistical mechanics, Ingber [67] developed an approach to elucidate the collective aspects of the neurocortical system *via nonlinear-nonequilibrium statistical mechanics*. In the relevant studies microscopic neural synaptic interactions consistent with anatomical observations were spatially averaged over columnar domains; and the relevant macroscopic spatial-temporal aspects were described by a *Lagrangian formalism* [68]. However, the topological constraints with the associated continuity relations posed by columnar domains and the Lagrangian approach are rather unrealistic.

5.7 Peretto's Model

A more pragmatic method of analyzing neural activity *via* statistical physics was portrayed by Peretto [38] who considered the collective properties of neural networks by extending Hopfield's model to Little's model. The underlying basis for Peretto's approach has the following considerations:

- Inasmuch as the statistical mechanics formalisms are arrived at in a Hamiltonian framework, Peretto "searches" for *extensive quantities* which depict the Hopfield network in the ground state as well as in noisy situations.

- Little's model introduces a markovian structure to neural dynamics. Hence, Peretto verifies whether the corresponding evolution equation would permit (at least with certain constraints) a Hamiltonian attribution to neural activity.

- Last, Peretto considers the feasibility of comparing both models in terms of the storage capacity and associative memory properties.

The common denominator in all the aforesaid considerations as conceived by Peretto again, is the statistical mechanics and/or spin-glass analogy that portrays a parallelism between Hopfield's network and Little's model of a neural assembly.

Regarding the first consideration, Peretto first enumerates the rules of synthesizing the Hopfield model, namely: 1) Every neuron i is associated with a membrane potential, V_i; 2) V_i is a linear function of the states of the neuron related to i or V_i refers to the somatic summation/integration given by $V_i = \sum_j C_{ij} S_j$ (with $S_j = 0$ or 1 according to the firing state of neuron i) and C_{ij} is the synaptic efficiency between the (upstream) neuron j and the (downstream) neuron i; 3) a threshold level V_{Ti} will decide the state of the neuron i as $S_i = +1$ if $V_i > V_{Ti}$ or as $S_i = 0$ if $V_i < V_{Ti}$.

Hence, Peretto develops the following Hamiltonian to depict the Hopfield neural model analogous to the Ising spin model:

$$H(I) = - \sum_{<ij>} J_{ij}' \sigma_i \sigma_j - \sum_i h_i{}^o \sigma_i \qquad (5.12)$$

where $\sigma_i = (2S_i - 1)$, (so that $\sigma_i = +1$ when $S_i = 1$ and $\sigma_i = -1$ when $S_i = 0$), $J_{ij} = C_{ij}/2$, $h_i^o = \Sigma_j C_{ij}/2 - V_{Ti}$ and $J_{ij} = (J_{ij} + J_{ji})$. In Equation (5.12), I represents the set of internal states namely, $I = \{\sigma_i\} = \{\sigma_1, \sigma_2, ..., \sigma_M\}$ for $i = 1, 2, ..., M$. The Hamiltonian H_N is identified as an extensive parameter of the system. (It should be noted here that the concept of Hamiltonian as applied to neural networks had already been proposed by Cowan as early as in 1967 [65]. He defined a Hamiltonian to find a corresponding invariant for the dynamics of a single two-cell loop.)

Concerning the second consideration, Peretto formulates an evolution equation to depict a Markov process. The relevant *master equation* written for the probability of finding the system state I at any time t, is shown to be a Boltzmann type equation and hence has a Gibbs' distribution as its steady-state solution.

Peretto shows that the markovian process having the above characteristics can be described by atleast a narrow class of Hamiltonians which obey the *detailed balance principle*. In other words, a Hamiltonian description of neural activity under markovian statistics is still feasible, though with a constraint posed by the detailed balance principle which translates to the synaptic interactions being symmetric ($J_{ij} = J_{ji}$).

Both Hopfield's model and Little's have been treated by Peretto under noisy conditions also. It is concluded that, in Hopfield's model, considering a Hebbian learning procedure* the Hamiltonian description of the neuronal state (analogous to that of the spin glass) can still be modified to ascertain the steady-state properties of the network exactly at any level of noise. (However, for a fully connected network, the dynamics is likely to become chaotic at times.)

Though the Hamiltonian approach of Little's model also permits the analysis of the network under noisy conditions, it is, however, more involved than Hopfield's model since it depends upon the noise level.

The last consideration, namely, the storage capacity of Hopfield's and Little's models by Peretto, leads to the inference that both models (which have the common basis *vis-a-vis* spin-glass analogy) present the same associative memory characteristics. There is, however, a small distinction: Little's model allows some *serial processing* (unlike Hopfield's model which

---

* *Hebbian learning* procedure here refers to unsupervised learning in which the synaptic strength (weight) is increased if both the source and destination neurons are activated. According to this learning rule, the synaptic strength is chosen as: $W_{ij} = (1/N) \Sigma^{PN}_{\mu=1} \sigma_i^\mu \sigma_j^\mu$, where N is the number of neurons of the network accommodating a storage of p_N patterns. Hebb's rule always leads to symmetric synaptic coupling.

represents a totally *parallel processing* activity). Hence, Peretto concludes that Little's model is more akin to biological systems.

Subsequent to Peretto's effort in compromising Hopfield's and Little's models on their behavior equated to the spin-glass system, Amit et al. in 1985 [69] analyzed the two dynamic models due to Hopfield and Little to account for the collective behavior of neural networks. Considering the long-time behavior of these models being governed by the statistical mechanics of infinite-range Ising spin-glass Hamiltonians, certain configurations of the spin system chosen at random are shown as memories stored in the *quenched random couplings*. The relevant analysis is restricted to a finite number of memorized spin configurations (patterns) in the thermodynamic limit of the number of neurons tending to infinity. Below the transition temperature (T_C) both models have been shown to exhibit identical long-term behavior. In the region $T < T_C$, the states, in general, are shown to be either metastable or stable. Below $T \cong 0.46 T_C$, dynamically stable states are assured. The metastable states are portrayed as due to mixing of the embedded patterns. Again, for $T < T_C$ the states are conceived as symmetric; and, in terms of memory configurations, the symmetrical states have equal overlap with several memories.

5.8 Little's Model *versus* Hopfield's Model

The Hopfield model defined by Equation. (5.12) with an associated memory has a well-defined dynamics. That is, given the initial pattern, the system evolves in time so as to relax to a final steady-state pattern. In the generalized Hopfield model* the transition probability $\rho(I/J)$ from state J to the next state I takes the usual form for $T > 0$ as:

$$\rho(I/J) = \exp\left[-\beta H_N(I)\right]/\Sigma \exp[-\beta H_N(K)] \tag{5.13}$$

and the system relaxes to the Gibbs distribution:

$$\rho(S_i) \cong \exp[-\beta H_N(S_i)] \tag{5.14}$$

In Little's model the transition probability is given by:

$$\rho(I/J) = \exp[-\beta H_N(I/J)]/\Sigma \exp[-\beta H_N(K/J)] \tag{5.15}$$

where

---

* In the original Hopfield model, a *single-spin flip* (*Glauber dynamics*) is assumed. This is equivalent to $T = 0$ in Monte-Carlo search procedure for the spin systems. The generalized model refers to $T > 0$.

$$H_N(I/J) = - \Sigma \, J_{ij} S_i(I) S_j(I) - \Sigma \, h_i S_i(I) \qquad (5.16)$$

Thus, in the Little model at each time-step, all the spins check simultaneously their states against the corresponding local field; and hence such an evolution is called *synchronous* in contrast with the Hopfield model which adopts the *asynchronous* dynamics.

Peretto has shown that Little's model leads to a Gibbs-type steady state exp $(-\beta H_N)$ where the effective Hamiltonian H_N is given by:

$$H_N(I/I) = H_N(I) = - (1/\beta) \, \Sigma \ln\left[\cosh\{\beta \, \Sigma \, J_{ij} S_j(I)\}\right]$$

$$(5.17)$$

This Hamiltonian specified by $H_N(I/I)$ corresponds to Hopfield's Hamiltonian of Equation (5.12).

The corresponding *free energy* of the Little model has been shown [33] to be twice that of the generalized Hopfield model at the extreme points. As a consequence, the nature of ground states and metastable states in the two models are identical as explained below.

Little [33] points out only a nontrivial difference between the neural network problem and the spin problem assuming a symmetry in the system. That is, as mentioned earlier, considering a matrix T_M consisting of the probabilities of obtaining a state $|S_1', S_2', ..., S_M'>$ given state $|S_1, S_2, ..., S_M>$ (where the primed set refers to the row and the unprimed set to the column of the element of the matrix) immediately preceding it, T_M is symmetric for the spin system. However, in neural transmission, the signals propagate from one neuron down its axon to the synaptic junction of the next neuron and not in the reverse direction; hence, T_M is clearly not symmetric for neural networks. That is, in general, the interaction of the jth neuron with the ith is not the same as that of the ith neuron with the jth.

Though T_M is not symmetric, Little [33] observes that the corresponding result can be generalized to an arbitrary matrix because, while a general matrix cannot always be diagonalized, it can, however, be reduced to so-called, *Jordan cannonical form* (see Appendix B); and Little develops the conditions for a persistent order based on the Jordan canonical form representation of T_M. Thus the asymmetry problem appears superficially to have been solved.

However, there are still many differences between physical realism and Little's model. The discrete time assumption as discussed by Little is probably the least physically acceptable aspect of both this model and the formal neuron. In addition, secondary effects such as potential decay, accommodation, and postinhibitory rebound are not taken into account in the model. To compare Little's model directly with real networks, details such as the synaptic connectivity should be known; and these can only be worked out only for a few networks. Thus, it should be emphasized that this model, like the formal neuron, represents only a minimal level of description of neural firing behavior.

As stated earlier, Thompson and Gibson [37] indicate that the spin-system definition of long-range order (fixing the spin at one lattice site causes the spins at sites far away from it to show a preference for one orientation) is not applicable to the neural problem. Contrary to Little [33], Thompson and Gibson state that the existence of order (a correlation between the probability distribution of the network at some initial time, and the probability distribution after m (m ≥ 1) time-steps does not mean that the network has a persistent state; and rather, order should only be considered over a moderate number of time-steps. However, inasmuch as order does imply a correlation between states of the network separated by time-steps, it seems reasonable to assume that order is associated with a memory mechanism.

Clearly, Little's model which is derived assuming a close similarity between it and the problem of an Ising system does not provide a comprehensive model of neural-firing behavior. However, it is advantageous in that the model neuron is both mathematically simple and able to produce a remarkably wide range of output patterns which are similar to the discharge patterns of many real neurons.

Further, considering Hopfield's model, Hopfield [31] states that for W_{ij} being symmetric and having a random character (analogous to the spin glass), state changes will continue until a local minimum is reached. That is, the equations of motions for a network with symmetric connections ($W_{ij} = W_{ji}$) always lead to a convergence to stable states in which the outputs of all neurons remain constant. Again, the symmetry of the network is essential to the mathematical description of the network. Hopfield notes that real neurons need not make synapses both of $i \to j$ and $j \to i$; and, without this symmetry, the probability of errors would increase in the input-output neural network simulation, and there is a possibility that the minimum reached *via* algorithmic search would only be metastable and could be replaced in time by another minimum. The question of symmetry and the symmetry condition can, however, be omitted without destroying the associative memory. Such simplification is justifiable *via* the principle of *universality* in physics which permits study of the collective aspects of a system's behavior by introducing separate (and more than one) simplifications without essentially altering the conclusions being reached.

The concept of memory or storage and retrieval of information pertinent to the Little and Hopfield models differ in the manner in which the state of the system is updated. In Little's model all neurons (spins) are updated synchronously as per the linear condition of output values, namely, $o_i(t) = \Sigma_j W_{ij} x_i(t)$, where the neurons are updated sequentially one at a time (either in a fixed order or randomly) in the Hopfield model. (Though sequential updating can be more easily simulated by conventional digital logic, real neurons do not operate sequentially.)

5.9 Ising Spin System *versus* Interacting Neurons

In view of the various models as discussed above, the considerations in the analogous representation of interacting neurons *vis-a-vis* the Ising magnetic spins and the contradictions or inconsistencies observed in such an analogy are summarized in Tables 5.1 and 5.2.

5.10 Liquid-Crystal Model

Basically, the analogy between Ising spins system and the neural complex stems from the fact that the organization of neurons is a collective enterprise in which the neuronal activity of interactive cells represents a cooperative process similar to that of spin interactions in a magnetic system. As summarized in Table 5.1, the strengths of synaptic connections between the cells representing the extent of interactive dynamics in the cellular automata are considered analogous to the strengths of exchange interactions in magnetic spin systems. Further, the synaptic activity, manifesting as the competition between the excitatory and inhibitory processes is regarded as equitable to the competition between the ferromagnetic and antiferromagnetic exchange interactions in spin-glass systems. Also, the threshold condition stipulated for the neuronal network is considered as the analog of the condition of metastability against single spin flips in the Ising spin-glass model.

Notwithstanding the fact that the aforesaid similarities do prevail between the neurons and the magnetic spins, major inconsistencies also, persist between these two systems regarding the synaptic coupling *versus* the spin interactions (Table 5.2). Mainly, the inconsistency between neurons with inherent asymmetric synaptic couplings and symmetric spin-glass interactions led Griffith [14] to declare the aggregate of neurons *versus* magnetic spin analogy as having "no practical value". Nevertheless, several compromising suggestions have been proposed as discussed earlier showing the usefulness of the analogy (between the neurons and the magnetic spins).

Table 5.1: Ising Spin System *versus* Neuronal System: Analogical Aspects

Magnetic Spin System	Neuronal System
Interacting magnetic spins represent a collective process.	Interacting neurons represent a collective process.
Dichotomous magnetic spin states: $\pm S_i$.	Dichotomous cellular potential states: $\sigma_i = 0$ or 1.
Exchange interactions are characterized by strengths of interaction.	Synaptic couplings are characterized by weights of synaptic connections.
Competition between ferromagnetic and antiferromagnetic exchange interactions.	Competition between the excitatory and inhibitory processes.
A set of M magnetic dipoles each with two spins ($\pm 1/2$).	A set of M neurons each with two potential states, 0 or 1.
Condition of metastability against single-spin flips.	Cellular state-transition crossing a threshold (metastable) state.
Phase transition from paramagnetism to ferromagnetism at a critical temperature (Curie point).	Onset of persistent firing patterns at a critical potential level.
A spin is flipped iff, the Hamiltonion (Lyapunov functional of energy) sets the dynamics of the spins to a ground-state.	A state of a neuron is changed iff, the Hamiltonion sets the dynamics of the neurons to converge to a local minimum (ground-state).

Table 5.2: Ising Spin System *versus* Neuronal System: Contradictions and Inconsistencies

Magnetic Spin System	Neuronal System
Microscopic *reversibility* pertaining to the magnetic spin interactions is inherent to the strength of coupling between the exchange interactions being symmetrical.	Symmetric weighting of neuronal interaction is questionable from the physiological viewpoint. This implies the prevalence of unequalness between the number of excitatory and inhibitory synapses.
That is, in the magnetic spin exchange interactions, the coupling coefficients $J_{ij} = J_{ji}$.	In the neuronal cycle of state-transitions, the interconnecting weights $W_{ij} \neq W_{ji}$.
The physical (molecular) arrangement of magnetic dipoles facilitates the aforesaid symmetry.	The physiological reality forbids the synaptic forward-backward symmetric coupling.
Symmetry in the state-transition matrix.	Asymmetry in the state-transition matrix .
Diagonalizable transition matrix.	Nondiagonalizable transition matrix.
No anisotropy in magnetic dipole orientations unless dictated by an external magnetic influence.	Anisotropy is rather inherent leading to a persistent order (in time as depicted by Little or in space as discussed in Section 5.10).
Hamiltonians obey the principle of detailed balance.	Only a subclass of Hamiltonians obey the principle of detailed balance.

The assumption of symmetry and the specific form of the synaptic coupling in a neuronal assembly define what is generally known as the Hopfield model. This model demonstrates the basic concepts and functioning of a neural network and serves as a starting point for a variety of models in which many of the underlying assumptions are relaxed to meet some of the requirements of real systems. For example, the question of W_{ij} being not equal to W_{ji} in a neural system was addressed in a proposal by Little (as detailed in the previous section), who defined a time-domain long-range order so that the corresponding anisotropy introduces bias terms in the Hamiltonian relation, making it asymmetric to match the neuronal Hamiltonian. That is, Little's long-range order as referred to neurons corresponds to a time-domain based long-time correlation of the states; and these persistent states (in time) of a neuronal network are equated to the long-range (spatial) order in an Ising spin system.

An alternative method of attributing the long-range order to neurons can be done by following the technique of Little except that such a long-range order will be referred to the spatial or orientational anisotropy instead of time correlations. To facilitate this approach, the *free-point molecular dipole interactions* can be considered *in lieu* of magnetic spin interactions [15]. The free-point molecular dipole interactions with partial anisotropy in spatial arrangement refer to the *nematic phase* in a *liquid crystal*. Hence, the relevant analysis equates the neural statistics to that of a nematic phase system consistent with the known dogma that "the living cell is actually a liquid crystal" [81]. That is, as Brown and Wolken [81] observed, the characteristics of molecular patterns, structural and behavioral properties of liquid crystals, make them unique model-systems to investigate a variety of biological phenomena. The general physioanatomical state of biological cells depicts neither real crystals nor real liquid phase (and constitutes what is popularly known as the *mesomorphous* state) much akin to several organic compounds which have become known as the *"flüssige Kristalle"* or liquid crystals; and both the liquid crystalline materials as well as the biological cells have a common, irregular pattern of side-by-side spatial arrangements in a series of layers (known as the *nematic phase*).

The microscopic structural studies of biological cells indicate that they are constituted by very complex systems of macromolecules which are organized into various bodies or *"organelles"* that perform specific functions for the cell. From the structural and functional point of view, Brown and Wolken have drawn an analogy of the description of the living cells to liquid crystals on the basis that a cell has a *structural order*. This in fact is a basic property of liquid crystals as well, for they have a structural order of a solid. Furthermore, in many respects it has been observed that the physical, chemical, structural, and optical properties of biological cells mimic closely those of liquid crystals.

Due to its liquid crystalline nature, a cell through its own structure forms a proto-organ facilitating electrical activity. Further, the anisotropically oriented structure of cellular assembly (analogous to liquid crystals) has been found responsible for the complex catalytic action needed to account for cellular regeneration. In other words, by nature the cells are inherently like liquid crystals with similar functional attributions.

On the basis of these considerations a neural cell can be modeled *via* liquid-crystal analogy, and the squashing action of the neural cells pertinent to the input-output relations (depicting the dynamics of the cellular automata) can be described in terms of a stochastically justifiable sigmoidal function and statistical mechanics considerations as presented in the pursuant sections.

5.11 Free-Point Molecular Dipole Interactions

Suppose a set of polarizable molecules are anisotropic with a spatial long-range orientational order corresponding to the nematic liquid crystal in the mesomorphic phase. This differs from the *isotropic molecular arrangement* (as in a liquid) in that the molecules are spontaneously oriented with their long axes approximately parallel. The preferred direction or orientational order may vary from point-to-point in the medium, but in the long-range, a specific orientational parallelism is retained.

In the nematic phase, the statistical aspects of dipole orientation in the presence of an externally applied field can be studied *via Langevin's theory* with the following hypotheses:

1. The molecules are point-dipoles with a prescribed extent of anisotropy.

2. The ensemble average taken at an instant is the same as the time average taken on any element (*ergodicity property*).

3. The characteristic quantum numbers of the problem are so high that the system obeys the classical statistics of Maxwell-Boltzmann, which is the limit of quantum statistics for systems with high quantum numbers. The present characterization of *paraelectricity* differs from spin *paramagnetism*, wherein the quantum levels are restricted to two values only.

4. The dipole molecules in general when subjected to an external electric field E, experience a moment $\mu_E = \alpha_E\,E$, where α_E by definition refers to the *polarizability* of the molecule. The dipole orientation contributing to the polarization of the

material is quantified as $P = N<\mu_E>$ where N is the dipole concentration.

5. In an anisotropic system such as the liquid crystal, there is a permanent dipole moment μ_{PE}, the direction of which is assumed along the long axis of a nonspherical dipole configuration. Consequently, two orthogonal polarizability components exist, namely, α_{E1} along the long axis and α_{E2} perpendicular to this long axis.

The dipole moments in an anisotropic molecule are depicted in Figure 5.1. Projecting along the applied electric field E the net-induced electric polarization moment is:

$$\mu_E = \mu_{PE}\cos\theta + (\alpha_{E1}\cos^2\theta + \alpha_{E2}\sin^2\theta)E$$

$$= \mu_{PE}\cos\theta + (\Delta\alpha_E\cos^2\theta + \alpha_{E2})E \qquad (5.18)$$

where $\Delta\alpha_E$ is a measure of anisotropy.

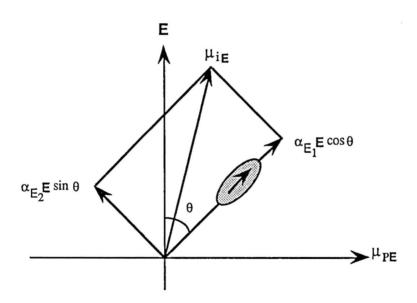

Figure 5.1 Free-point dipole and its moments
E: Applied electric field; μ_{PE}: Permanent dipole moment;
μ_{iE}: Induced dipole moment

The corresponding energy of the polarized molecule in the presence of an applied field E is constituted by: (1) The potential energy W_{PE} due to the permanent dipole given by,

$$W_{PE} = -\mu_{PE} \cdot E = -\mu_{PE} E \cos \theta \qquad (5.19)$$

and (2) the potential energy due to the induced dipole given by:

$$W_{iE} = -(1/2)(\alpha_{E1}\cos^2\theta + \alpha_{E2}\sin^2\theta)|E|^2 \qquad (5.20)$$

Hence, the total energy is equal to $W_T = W_{PE} + W_{iE}$. Further, the statistical average of μ_E can be specified by:

$$<\mu_E> = \frac{\int \mu_E \exp[-W_{PE}/k_BT]d\Omega}{\int \exp[-W_T/k_BT]d\Omega} \qquad (5.21)$$

where $d\Omega$ is the elemental solid angle around the direction of E. That is, $d\Omega = 2\pi \sin(\theta)d\theta$. By performing the integration of Equation (5.21) using Equation (5.18), it follows that:

$$<\mu_E> = \mu_{PE}<\cos\theta> + (\Delta\alpha_E<\cos^2\theta> + \alpha_{E_2})E \qquad (5.22)$$

where the quantity $<\cos^2\theta>$ varies from $1/3$ (for randomly oriented molecules) to 1 for the case where all the molecules are parallel (or antiparallel) to the field E. On the basis of the limits specified by $<\cos^2\theta>$, the following parameter can be defined:

$$\begin{aligned}
S^o \quad &= (3/2) <\cos^2\theta> - (1/2) \\
&= 0 \quad (\text{for } <\cos^2\theta> = 1/3) \\
&= 1 \quad (\text{for } <\cos^2\theta> = 1)
\end{aligned} \qquad (5.23)$$

The parameter S^o which is bounded between 0 and 1 under the above conditions, represents the "order parameter" of the system [82]. Appropriate to the nematic phase, S^o specifies the long-range orientational parameter pertaining to a liquid crystal of rod-like molecules as follows: Assuming the distribution function of the molecules to be cylindrically symmetric about the axis of preferred orientation, S^o defines the degree of alignment, namely, for perfectly parallel (or antiparallel) alignment $S^o = 1$, while for random orientations $S^o = 0$. In the nematic phase S^o has an intermediate value which is strongly temperature dependent.

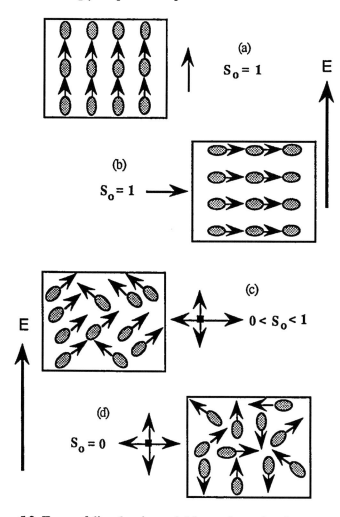

Figure 5.2 Types of disorders in spatial free-point molecular arrangement subjected to external electric field (**E**)
(a) & (b) Completely ordered (total anisotropy): Parallel and antiparallel arrangements; (c) Partial long-range order (partial anisotropy): Nematic phase arrangement; (d) Complete absence of long-range order (total isotropy): Random arrangement

For $S^o = 0$, it refers to an isotropic statistical arrangement of random orientations so that for each dipole pointing in one direction, there is statistically a corresponding molecule in the opposite direction

(Figure 5.2). In the presence of an external electric field **E**, the dipoles experience a torque and tend to polarize along **E**, so that the system becomes slightly anisotropic; and eventually under a strong field (**E**) the system becomes totally anisotropic with $S^o = 1$.

5.12 Stochastical Response of Neurons under Activation

By considering the neurons as analogous to a random, statistically isotropic dipole system, the graded response of the neurons under activation could be modeled by applying the concepts of Langevin's theory of dipole polarization; and the continuous graded response of neuron activity corresponding to the stochastical interaction between incoming excitations that produce true, collective, nonlinear effects can be elucidated in terms of a sigmoidal function specified by a gain parameter $\lambda = \Lambda/k_B T$, with Λ being the scaling factor of σ_i which depicts the neuronal state-vector.

In the pertinent considerations, the neurons are depicted similar to the nematic phase of liquid crystals and are assumed to possess an inherent, long-range spatial order. In other words, it is suggested that $0 < S^o < 1$ is an appropriate and valid *order function* for the neural complex that $S^o = 0$. Specifying in terms of $S^o = (3/2) <\cos^2\theta> - 1/2$, the term $<\cos^2\theta>$ should correspond to a value between 1/3 to 1 (justifying the spatial anisotropy).

To determine an appropriate squashing function for this range of $<\cos^2\theta>$ between 1/3 to 1 (or for $0 < S^o < 1$), the quantity $<\cos^2\theta>$ can be replaced by $(1/3 + 1/3q)$ in defining the order parameter S^o. Hence:

$$S^o = (3/2)(1/3 + 1/3q) - (1/2) \tag{5.24}$$

where $q \to \infty$ and $q = 1/2$ set the corresponding limits of $S^o = 0$ and $S^o = 1$ respectively.

Again, resorting to statistical mechanics, $q = 1/2$ refers to dichotomous states, if the number of states are specified by $(2q + 1)$. For the dipoles or neuronal alignments, it corresponds to the two totally discrete anisotropic (parallel or antiparallel) orientations. In a statistically isotropic, randomly oriented system, the number of (possible) discrete alignments would, however, approach infinity, as dictated by $q \to \infty$.

For the intermediate $(2q + 1)$ number of discrete orientations, the extent of dipole alignment to an external field or, correspondingly, the (output) response of a neuron to excitation would be decided by the probability of a discrete orientation being realized. It can be specified by [83]:

$$L_q(x) = \sum_{m=-q}^{+q}(m/q)\exp(mx/q) \Big/ \sum_{m=-q}^{+q}\exp(mx/q)$$

$$= (1 + 1/2q)\coth[(1 + 1/2q)x] - (1/2q)\coth(x/2q) \tag{5.25}$$

The above function, $L_q(x)$, is a *modified Langevin function* and is also known as the *Bernoulli function*. The traditional Langevin function $L(x)$ is the limit of $L_q(x)$ for $q \to \infty$. The other limiting case, namely, $q = 1/2$, which exists for dichotomous states, corresponds to $L_{1/2}(x) = \tanh(x)$.

Thus, the sigmoidal function $F_S(x)$ which decides the neuronal output response to an excitation has two bounds: With $F_S(x) = \tanh(x)$, it corresponds to the assumption that there exists a total orientational long-range order in the neuronal arrangement. Conventionally [16], $F_S(x) = \tanh(x)$ has been regarded as the *squashing function* (for neuronal nets) purely on empirical considerations of the input-output nonlinear relation being S-shaped (which remains bounded between two logistic limits, and follows a continuous monotonic functional form between these limits). In terms of the input variate x_i and the gain/scaling parameter Λ of an ith neuron, the sigmoidal function specified as the hyberbolic tangent function is $\tanh(\Lambda x_i)$. The logistic operation that compresses the range of the input so that the output remains bounded between the logical limits can also be specified alternatively by an exponential form, $F_S(y) = 1/[(1 + \exp(-y)]$ with $y = \Lambda x_i$.

Except for being sigmoidal, the adoption of the hyperbolic tangent or the exponential form in the neural network analyses has been purely empirical with no justifiable reasoning attributed to their choice. Pursuant to the earlier discussion, $L(y) = L_{q \to \infty}(y)$ specifies the system in which the randomness is totally isotropic. That is, the anisotropicity being zero is implicit. This, however, refers to rather an extensive situation assuming that the neuronal configuration poses no spatial anisotropicity or long-range order whatsoever. Likewise, considering the intuitive modeling of $F_S(y) = \tanh(y)$, as adopted commonly, it depicts a totally anisotropic system wherein the long-range order attains a value one. That is, $\tanh(y) = L_{q \to 1/2}(y)$ corresponds to the dichotomous discrete orientations (parallel or antiparallel) specified by $(2q + 1) \to 2$.

In the nematic phase, neither of the above functions, namely, $\tanh(y)$ nor $L(y)$, is commensurable since a partial long-range order (depicting a partial anisotropicity) is rather imminent in such systems. Thus, with $1/2 < q < \infty$, the true sigmoid of a neuronal arrangement (with an inherent nematic, spatial long-range order) should be $L_q(y)$.

Therefore, it can be regarded that the conventional sigmoid, namely, the hyperbolic tangent (or its variations) and the Langevin function, constitute the upper and lower bounds of the state-vector squashing characteristics of a neuronal unit, respectively.

Relevant to the above discussions, the pertinent results are summarized in Table 5.3.

Table 5.3: Types of Spatial Disorder in the Neural Configuration

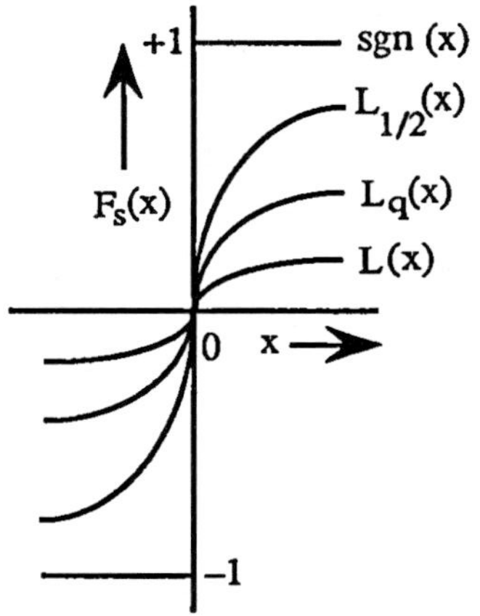

Figure 5.3 Sigmoidal function

Sigmoids $F_S(x)$: sgn(x), $L_{1/2}(x)$, $L_q(x)$, $L(x)$

* (Analogous to quantum numbers)

Neural Spatial Configuration	Sigmoidal Function $F_S(x)$	Slope at the Orgin $dF_S(x)/dx\|_{x=0}$ (m_o)	Order Parameter S^o	Remark	Number of Orientational States*
Totally Isotropic (Zero Long-Range Order)	Langevin Function $L_q(x) = L(x)$ $q \to \infty$	1/3	0	Lower Bound Sigmoid	∞ $(q \to \infty)$
Totally Anisotropic (Long-Range Order to Total Extent)	Hyperbolic Tangent $L_q(x)$ $q \to 1/2$	1	1	Upper Bound Sigmoid	2 $(q \to 1/2)$
Partially Anisotropic (Partial Long-Range Order as in Nematic Phase)	Bernoulli Function $L_q(x)$ $1/2 < q < \infty$	$[1/3 + 1/3q]$	$3/2(m_o) - 1/2$ $(0 < S^o < 1)$	Nematic Phase Neuronal Configuration with Partial Long-Range Order	$(2q+1)$
Zero Interaction	Signum Function $L_q(x)$ $q \to 0$	∞	—	McCulloch-Pitts' Model	—

5.13 Hamiltonian of Neural Spatial Long-Range Order

In general, the anisotropicity of a disorder leads to a Hamiltonian which can be specified in two ways: (1) Suppose the exchange Hamiltonian is given by:

$$H_N = - \Sigma \ W_{xx}S_i^xS_j^x + W_{yy}S_i^yS_j^y + W_{zz}S_i^zS_j^z) \qquad (5.26)$$

where W_{xx}, W_{yy} and W_{zz} are diagonal elements of the exchange matrix W (with the off- diagonal elements being zero). If $W_{xx} = W_{yy} = 0$ and $W_{zz} \neq 0$, it is a symmetric anisotropy (with dichotomous states as in the Ising model). Note that the anisotropy arises if the strength of at least one of the exchange constants is different from the other two. If $W_{xx} = W_{yy} \neq 0$ and $W_{zz} = 0$, it corresponds to an isotropic xy model; and, if $W_{xx} = W_{yy} = W_{zz}$, it is known as the *isotropic Heisenberg model*. (2) Given that the system has an anisotropy due to partial long-range order as in the nematic phase representation of the neuronal arrangement, the corresponding Hamiltonian is:

$$H_N = - W\Sigma \ S_iS_j + H_a \qquad (5.27)$$

where H_a refers to the anisotropic contribution which can be specified by an inherent constant h_i^o related to the order parameter, S^o, so that

$$H_N = -\underset{i}{\Sigma} \ \underset{j}{\Sigma} \ W_{ij}S_iS_j - \underset{i}{\Sigma} \ h_i^oS_i \qquad (5.28)$$

While the interactions W_{ij} are local, H_N refers to an extensive quantity corresponding to the long-range orientational (spatial) interconnections in the neuronal arrangement.

5.14 Spatial Persistence in the Nematic Phase

The nematic-phase modeling of the neuronal arrangement specifies (as discussed earlier) a long-range spatial anisotropy which may pose a persistency (or preferred, directional routing) of the synaptic transmission. Pertinent analysis would be similar to the time-domain persistency demonstrated by Little [33] as existing in neuronal firing patterns.

Considering $(2q + 1)$ possible spatial orientations (or states) pertaining to M interacting neurons as represented by $\Psi(\alpha)$, then the probability of obtaining the state $\Psi(\alpha')$, having started with a preceding $\Psi(\alpha)m$ spatial intervals {x}, can be written in terms of a transfer matrix T_M^m as:

105

$$\Psi(\alpha') \, T_M^m \, \Psi(\alpha) \tag{5.29}$$

where $\Psi(\alpha)$ can be expressed in terms of $(2q + 1)^M$ orthonormal eigenvectors ϑ_r (with eigenvalues λ_r) of the operator T_M. Each ϑ_r has $(2q + 1)^M$ components, one for each configuration α; that is:

$$\Psi(\alpha) = \sum_{r=1}^{(2q+1)^M} \vartheta_r(\alpha) \tag{5.30}$$

Hence

$$\langle \Psi(\alpha') | T_M | \Psi(\alpha) \rangle = \sum_r \lambda_r \vartheta_r(\alpha') \vartheta_r(\alpha) \tag{5.31}$$

Analogous to the time-domain persistent order analysis due to Little, it is of interest to find a particular state α_1 after m spatial steps, having started at an arbitrary commencement (spatial location) in the neuronal topology; and hence the probability of obtaining the state α_2 after ℓ spatial steps, given α_1 after m spatial steps from the commencement location, can be written as:

$$\Gamma(\alpha_1, \alpha_2) = \Gamma(\alpha_1)\Gamma(\alpha_2) \tag{5.32}$$

which explicitly specifies no spatial correlation between the states α_1 and α_2. However, if the maximum eigenvalue λ_{max} is degenerate, the above factorization of $\Gamma(\alpha_1, \alpha_2)$ is not possible and there will be a spatial correlation in the synaptic transmission behavior. Such a degeneracy (in spatial order) can be attributed to any possible transition from isotropic to anisotropic nematic phase in the neuronal configuration. That is, in the path of synaptic transmission, should there be a persistent or orientational linkage/interaction of neurons, the degeneracy may automatically set in. In the spin system, a similar degeneracy refers to the transition from a paramagnetic to ferromagnetic phase. In a neural system, considering the persistency in the time-domain, Little [33] observes that long-range time-ordering is related to the short-term memory considerations as dictated by intracellular biochemical process(es).

5.15 Langevin Machine

The integrated effect of all the excitatory and inhibitory postsynaptic axon potentials in a neuronal network, which decides the state transition (or "firing"), is modeled conventionally by a network with multi-input state-vectors S_i (i = 1, 2, ..., M) with corresponding outputs σ_j (j = 1, 2, ..., N) linked *via* N-component weight-states W_{ij} and decided by a nonlinear activation function. The corresponding input-output relation is specified by:

$$S_i = \sum_{j=1}^{N} W_{ij} \sigma_j + \theta_i + \xi_n \qquad (5.33)$$

where θ_i is an external (constant) bias parameter that may exist at the input and ξ_n is the error (per unit time) due to the inevitable presence of noise in the cellular regions.

The input signal is further processed by a nonlinear activation function F_S to produce the neuron's output signal, σ. That is, each neuron randomly and asynchronously evaluates its inputs and readjusts σ_i accordingly.

The justification for the above modeling is based on Hopfield's [31,36] contention that real neurons have continuous, monotonic input-output relations and integrative time-delays. That is, neurons have sigmoid (meaning S-shaped) input-output curves of finite steepness rather than the steplike, two-state response curve or the logical neuron model suggested by McCulloch and Pitts [7].

The commonly used activation function to depict the neuronal response as mentioned earlier is the hyperbolic tangent given by $F_S(\Lambda\sigma_i) = \tanh(\Lambda\sigma_i)$ where Λ is a gain/scaling parameter. It may be noted that as Λ tends to infinity, $F_S(\Lambda\sigma_i)$ becomes the signum function indicating the "all-or-none" response of the McCulloch-Pitts model.

Stornetta and Huberman [84] have noted about the training characteristics of back-propagation networks that the conventional 0-to-1 dynamic range of inputs and hidden neuron outputs is not optimum. The reason for this surmise is that the magnitude of weight adjustment is proportional to the output level of the neuron. Therefore, a level of 0 results in no weight modification. With binary input vectors, half the inputs, on the average, will, however, be zero and the weights they connect to will not train. This problem is solved by changing the input range to $\pm 1/2$ and adding a bias to the squashing function to modify the neuron output range to $\pm 1/2$. The corresponding squashing function is as follows:

$$F_S(x) = -1/2 + 1/[1 + \exp(-x)] \qquad (5.34)$$

which is again akin to the hyperbolic tangent and/or exponential function forms discussed earlier.

These aforementioned sigmoids are symmetrical about the origin and have bipolar limiting values. They were chosen on an empirical basis, purely on the considerations of being S-shaped. That is, by observation, they match Hopfield's model in that the output variable for the ith neuron is a continuous and monotone-increasing function of the instantaneous input to i^{th} neuron having bipolar limits.

In Section 5.12, however, the Langevin function has been shown to be the justifiable sigmoid on the basis of stochastical attributions of neuronal activity and the implications of using Langevin function *in lieu of* the conventional sigmoid in the machine description of neuronal activity are discussed in the following section. Such a machine is designated as the *Langevin machine.*

5.16 Langevin Machine *versus* Boltzmann Machine

In Boltzmann machines, the neurons change state in a statistical rather than a deterministic fashion. That is, these machines function by adjusting the states of the units (neurons) asynchronously and stochastically so as to minimize the global energy. The presence of noise is used to escape from the local minima. That is, as discussed in Chapter 4, occasional (random) jumps to configurations of higher energy are allowed so that the problem of stabilizing to a local rather than a global minimum (as suffered by Hopfield nets) is largely avoided. The Boltzmann machine rule of activation is decided probabilistically so that the output value σ_i is set to one with the probability $p(\sigma_i = 1)$, where p is given in Equation (4.8), regardless of the current state. As discussed in Chapter 4 Akiyama et al. [54] point out that the Boltzmann machine corresponds to the Gaussian machine in that the sigmoidal characteristics of p fit very well to the conventional gaussian cumulative distribution with identical slope at the input $\sigma_i = 0$ with an appropriate choice of the scaling parameter.

The Boltzmann function, namely, $\{1/[1 + \exp(-x)]\}$ and the generalized Langevin function $[1 + L_q(x)]/2$ represent identical curves with a slope of $+1/4$ at $x = 0$, if q is taken as $+4$. Therefore, inasmuch as the Boltzmann machine can be matched to the gaussian machine, the Langevin function can also be matched likewise; and, in which case, it is termed here as the *Langevin machine.*

Considering neural network optimization problems, sharpening schedules which refer to the changes in the reference level with time adaptively are employed in order to achieve better search methods. Such a scheduling scheme using Langevin machine strategies is also possible and can be expressed as:

$$a_0 = A_0[1 - L(t/\tau_{a0})] \qquad (5.35)$$

where a_0 is the reference activation level which is required to decrease over time. A_0 is the initial value of a_0, and τ_{a0} is the time constant of the sharpening schedule.

Using the Langevin machine, the annealing can also be implemented by the following scheme:

$$T = T_0[1 - L_q(t/\tau_{Tn})] \tag{5.36}$$

where T_0 is the initial temperature and τ_{Tn} is the time constant of the annealing schedule which may differ from τ_{a0}. By proper choice of q, the speed of annealing can be controlled.

5.17 Concluding Remarks

The formal theory of stochastic neural network is based heavily on statistical mechanics considerations. However, when a one-to-one matching between real neuronal configuration and stochastical neural networks (evolved from the principles of stochastical mechanics) is done, it is evident that there are as many contradictions and inconsistencies as the analogies that prevail in such a comparison. The analogies are built on the common notion, namely, the interactive collective behavior of the ensemble of units — the cells in the neural complex and the magnetic spins in the material lattice. The inconsistencies blossom from the asymmetric synaptic coupling of the real neurons as against the inherently symmetric attributes of magnetic spin connection strengths.

Hopfield's ingenious "step backward" strategy of incorporating a symmetry in his neural models and the *pros and cons* discussions and deliberations by Little and Perelto still, however, dwell on the wealth of theoretical considerations pertinent to statistical mechanics consistent with the fact that a "reasearch under a paradigm must be a particularly effective way of inducing a paradigm change".

CHAPTER 6

Stochastical Dynamics of the Neural Complex

6.1 Introduction

The integrated effect of all the excitatory and inhibitory postsynaptic axon potentials in the neural complex (Figure 6.1) which decide the state transition (or "firing") is modeled conventionally by a network with multi-input state-vectors S_i (i = 1, 2, ..., M) with corresponding outputs σ_j (j = 1, 2, ..., N) linked *via* N-component weight-states W_{ij} and decided by a nonlinear activation function. That is, as indicated earlier:

$$S_i = \sum_{j=1}^{N} W_{ij}\, \sigma_j + \theta_i + e_n \tag{6.1}$$

where θ_i is an external (constant) bias parameter that may exist at the input and e_n is the error (per unit time) due to the presence of intra- or extra-neural disturbances. This unspecified noise source permits invariably the neurons to change their internal states in a random manner. The resulting error would upset the underlying learning or training process that sets the weighting vector W_{ij} to such a position as to enable the network to achieve maximization (or minimization) in its global (output) performance (such as mean-squared error). The corresponding *stability dynamics* of the neural activity can be specified by a *nonlinear stochastical equation* governing the variable, namely, the weighting vector W, as described in the following sections [18].

6.2 Stochastical Dynamics of the Neural Assembly

The state-transition in a neural complex (in its canonical form) represents a dichotomous (bistable) process and the presence of noise would place the bistable potential at an unstable equilibrium point. Though the initial (random) fluctuations/disturbances can be regarded as microscopic variables, with the progress of time (in an intermediate time domain), the fluctuation enhancement would be of macroscopic order. Such fluctuations can be specified in general by nonlinear stochastical dynamics governed by a relaxational *Langevin equation*, namely [85]:

$$\partial W/\partial t = \gamma\, C(W) + \eta(t) \tag{6.2}$$

where γ is a positive coefficient and $C(W)$ is an arbitrary function representing possible constraints on the range of weights with a parameter of nonlinearity; and $\eta(t)$ represents the driving random disturbance usually regarded as zero mean, gaussian white noise with a variance equal to

$\langle\eta(t)\eta(t')\rangle = 2k_BT\delta(t - t')$, where k_BT represents the pseudo-thermodynamic Boltzmann energy.

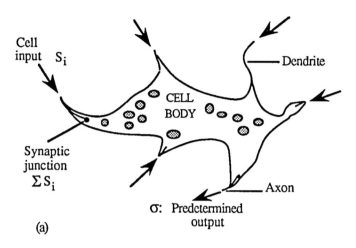

(a)

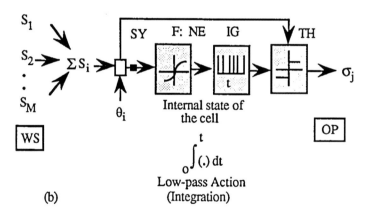

(b)

Figure 6.1 The biological neuronal cell structure and its network equivalent
(a) Biological neuron; (b) Network equivalent of the cell
SY: Synapse; F: Sigmoid; IG: Impulse generator; NE: Nonlinear estimator;
WS: Weighted sum of external inputs and weighted input from other
neurons; θ_i: External bias; TH: Threshold; OP: Output

Equation (6.2) can be specified alternatively by an equivalent *Fokker-Planck relation* depicting the probability distribution function $P(W, t)$, given by [86,87]:

111

$$\partial P(W, t)/\partial t = [-\gamma\, \partial C(W)/\partial W]P(W, t) + k_B T \partial^2 P(W, t)/\partial W^2$$
$$(6.3)$$

The stable states of the above equation are decided by the two extrema of the variable W, namely, $\pm W_m$; and the unstable steady state corresponds to $W_m \to 0$. The evolution of $W(t)$ or $P(W, t)$ depends critically on the choice of initial conditions with two possible modes of fluctuations: When the mean-squared value of the fluctuations is much larger than $k_B T$, it refers to an *extensive regime* depicting the passage of the states from apparently unstable to a preferred stable state. This is a *slow-time evolution process*. The second category corresponds to the mean-squared value being much smaller than $k_B T$ which specifies an intrinsic unstable state; and it depicts the evolution of the neuronal state from an unstable condition to two steady states of the McCulloch-Pitts regime [7]. For neuronal state disturbances, the relevant evolution process fits, therefore, more closely to the second type [33].

Invariably, the intrinsic fluctuations are correlated in time; and in view of the *central limit theorem*, they may also be gaussian. Further, the spectral characteristics of this noise are be *band-limited* (*colored*) as justified in the next section.

Hence, the solution of the Langevin equation (6.2) and/or the Fokker-Planck equation (6.3) depicting the state-transition behavior of the neural network should in general, refer to the fluctuations being a gaussian-colored noise causing the action potentials to recur randomly with a finite correlation time and thereby have a markovian structure as described below.

6.3 Correlation of Neuronal State Disturbances

The statistical aspects of random intervals between the action potentials of biological neurons are normally decided by the irregularities due to neural conduction velocity/dynamics, axonal fiber type mixture, synchronization/asynchronization effects (arising from dropping out of certain neuron units in the synaptic transmission), percentage of polyphasic action potentials, etc. The temporal dynamics of the neural conduction (or the action potential) can therefore be modeled as a train of delta-Dirac functions (representing a symmetric dichotomous process with bistable values), the time interval between their occurrences being a random variable (Figure 6.2a). Though in a *memoryless* mathematical neuron model, the relevant statistics of the recurrence of action potentials is presumed to be independent of the other events; the process underlying the neuronal disturbance cannot be altogether assumed as free of dependency on the previous history. As pointed out by McCulloch and Pitts [7], there is a possibility that a particular neural state has dependency at least on the preceding event.

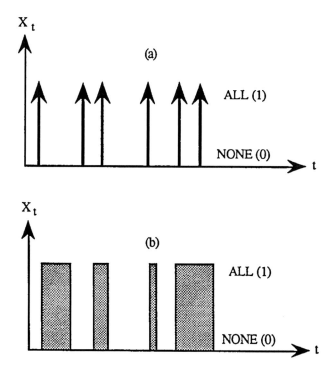

Figure 6.2 Models of action potential train
(a) Delta-dirac (impulse) representation; (b) Semirandom telegraphic signal
representation (markovian statistics)

In other words, markovian statistics can be attributed to the neuronal state transition and the occurrence of action potentials can be modeled as a symmetric dichotomous Markov process which has bistable values at random intervals. The *waiting times* in each state are exponentially distributed (which ensures markovian structure of the process involved) having a correlation function given by:

$$<X_t X_{t+\tau}> = \rho^2 \exp(-\Gamma\tau) \tag{6.4}$$

Here, $\rho^2 = (k_B T\Gamma)$ and the symmetric dichotomous Markov variable X_t represents the random process whose value switches between two extremes (all-or-none) $\pm\rho$, at random times. The correlation time τ_c is equal to $1/\Gamma$, and the mean frequency of transition from one value to the other is $\Gamma/2$. That is, the stochastic system has two state epochs, namely, random intervals of occurrence and random finite duration of the occurrences. (In a simple delta-Dirac representation, however, the durations of the disturbances are assumed to be of zero values). The process X_t ($t \geq 0$) should therefore represent approximately a *semirandom telegraphic* signal (Figure 6.2b).

The transition probability between the bistable values is dictated by the *Chapman-Kolmogorov system of equations* [88] for integer-valued variates; and the spectral density of the dichotomous Markov process can be specified by the Fourier transform of the correlation function given by Equation (6.4) and corresponds to the well-known *Lorentzian relation* given by:

$$S_\tau(\omega) = \Gamma \rho^2 / \pi (\omega^2 + \Gamma^2) \tag{6.5}$$

It can be presumed that a synchronism exists between the recursive disturbances, at least on a short-term/quasistationary basis [33]. This could result from more or less simultaneous activation of different sections of presynaptic fibers.

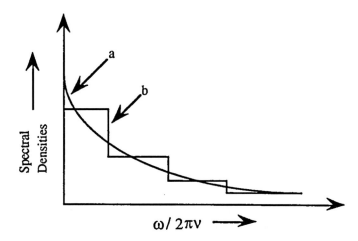

Figure 6.3 Spectral densities of a dichotomous markov process
(a): Under aperiodic limit; (b): Under periodic limit

In the delta-Dirac function model, this synchronism is rather absolute and implicit. In the markovian dichotomous model, the synchronism can be inculcated by a periodic attribute or an external parameter T_S so that the periodic variation will mimic the dichotomous Markov process with a correlation time $1/\Gamma$, assuming average switching frequencies of the two variations to be identical. That is, $2/T_S = \Gamma/2$. For this periodic fluctuation, the correlation function is a sawtooth wave between $\pm \rho^2$ of fundamental angular frequency $2\pi\nu$. The corresponding Fourier spectrum is given by:

$$S_{T_s}(\omega) = \sum_q [16\rho^2 v^2/\omega^2] \, \delta(\omega - 2\pi qv) \qquad\qquad q = \pm 1, \pm 3, ..., \infty$$

$$(6.6)$$

where $\delta(\omega - 2\pi qv)$ is an impulse unit area occurring at frequency $\omega = 2\pi qv$. Typical normalized spectral densities corresponding to the periodic and aperiodic limits of dichotomous Markov process are depicted in Figure 6.3.

Inasmuch as the output of the neuronal unit has a characteristic colored noise spectrum, it can be surmised that this limited bandwidth of the noise observed at the output should be due to the intrinsic, nonwhite spectral properties of the neuronal disturbances. This is because the firing action at the cell itself would not introduce band-limiting on the intrinsic disturbances. The reason is as follows: The state changing or the time response of the neuronal cell refers to a signum-type switching function or a transient time response as depicted in Figure 6.4.

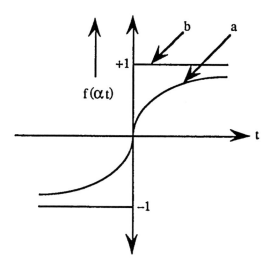

Figure 6.4 Transient response of a neuronal cell
(a): For arbitrary values of $\alpha < \infty$; (b): For the value of $\alpha \to \infty$

The transient time response $f(\alpha t)$ has a frequency spectrum specified in terms of the Laplace transform given by:

$$S(\omega) = \mathcal{L}[f(\alpha t)] = |-(2/\omega) - (4/\alpha^2) \sum_{q=1}^{\infty} \omega/[(2q)^2 + (\omega/\alpha)^2]$$
$$+ (\pi/\alpha) \, \text{sign} \, (\omega/\alpha)| \qquad\qquad (6.7)$$

where α is a constant and as $\alpha \to \infty$, the transient response assumes the ideal signum-type switching function in which case the frequency is directly proportional to $(1/\omega)$. However, the output of the neuronal unit has $(1/\omega^2)$ spectral characteristics as could be evinced from Equation (6.5). Therefore, the switching action at the neuronal cell has less influence in dictating the output spectral properties of the noise. In other words, the colored frequency response of the disturbances elucidated at the output should be essentially due to the colored intrinsic/inherent spectral characteristics of the disturbances existing at the neuronal structure.

Hence, in general, it is not justifiable to presume the spectral characteristics of neural disturbances as a flat-band white noise and the solutions of Langevin and/or Fokker-Planck equations described before should therefore correspond to a colored noise situation.

6.4 Fokker-Planck Equation of Neural Dynamics

The state of neural dynamics as indicated earlier is essentially decided by the intrinsic disturbances (noise) associated with the weighting function **W**. Due to the finite correlation time involved, the disturbance has band-limited (colored) gaussian statistics.

Ideally in repeated neuronal cells, there could be no coherence between state transitions induced by the disturbance/noise, or even between successive transitions. Such complete decorrelation is valid only if the noise or disturbance level is very small. However, inasmuch as the correlation does persist, the state-variable **W** specified before in an M-dimensional space $(W_i, i = 1, 2, 3, ..., M)$, can be modeled as a simple version of Equation (6.3). It is given by [89]:

$$\partial W(t)/\partial t = W(t) - [W(t)]^3 + \eta(t) \qquad (6.8)$$

where $\eta(t)$ is the noise term such that $<\eta(t)> = 0$ and $<\eta(t)\eta'(t)> = (k_B T)\Gamma \exp\{-\Gamma|t - t'|\}$ where $\Gamma \to \infty$ sets the limit that the above Fokker-Planck relation corresponds to the white noise case.

In the state-transition process, the relevant instability dynamics can be dictated by a set of stochastical differential equations, namely:

$$\partial W_i/\partial t = D_i(W) + \eta_i(t) \qquad (6.9)$$

where $\eta_i(t)$ refers to the noise/disturbance involved at the ith cell and $D_i(W)$ is an arbitrary function of **W**. Here as **W** approaches a representative value, say W_0, at $t = 0$ so that $D_i(W_0) = 0$, then the state-transition is regarded as unstable. An approximate solution to Equation. (6.9) can then be sought by assuming that $P(W, 0) = \delta(W - W_0)$ as the initial condition. This can be done by the *scaling procedure* outlined by Valsakumar [89]. Corresponding to Equation (6.9), a new stochastic process defined by a variable $\zeta(t)$ can be conceived such that in the limit of vanishing noise Equation (6.9) would refer to the new variable $\zeta(t)$, replacing the original variable $W(t)$. The correspondence between $\zeta(t)$ and $W(t)$ can then be written as [89]:

$$\partial \zeta_i(t)/\partial t = \sum_{j=1}^{N} [\partial \zeta_i(t)/\partial W_j] \, \eta_j \qquad (6.10)$$

At the enunciation of instability (at $t = 0$), the extent of disturbance/noise is important; however, as the time progresses, the nonlinearity associated with the neuronal state transition overwhelms. Therefore, the initial fluctuations can be specified by replacing $\partial \zeta(t)/\partial W_j$ in Equation (6.10) by its value at the unstable point. This refers to a *scaling approximation* and is explicitly written as:

$$(\partial \zeta_i/\partial W_j)|_{scaled} \approx (\partial \zeta_i/\partial W_j)|_{W = W_0} \qquad (6.11)$$

The above approximation leads to a correspondence relation between the probability distribution $P_{\zeta_s}(\zeta_s, t)$ of the scaled variable ζ_s and distribution function $P_W(W, t)$. The scaling solution to Equation (6.8) is hence obtained as [89]:

$$P_{\zeta_s}(\zeta_s, t) = [1/(2\pi\beta)^{1/2}] \exp(-\zeta_s^2/2\beta) \qquad (6.12a)$$

where

$$\beta = \{[k_B T \Gamma/(\Gamma^2 - 1)] \{\Gamma[1 - \exp(-2t)]$$
$$- [1 + \exp(-2t) - 2\exp(-1-\Gamma)t]\} \qquad \text{for } \Gamma \neq 1$$

$$= (k_B T/2) [1 - \exp(-2t) - (2t)\exp(-2t)] \qquad \text{for } \Gamma = 1$$

$$(6.12b)$$

The various moments (under scaling approximation) are:

$$<W^{2m+1}>_s = 0 \qquad (6.13)$$

$$<W^{2m}>_s = \{1/[1 - \exp(-2t)]^m\}(\pi)^{-1/2} \int_{-\infty}^{\infty} (u^2\tau'/[1 + u^2\tau'])^m \exp(-u^2) \, du$$

$$(6.14)$$

where $[\tau'/T_d(\Gamma)] = 2 \, (\beta/k_B T)[\exp(2t) - 1]$ and $T_d(\Gamma)$ is the *switching-delay* given by:

$$T_d(\Gamma) = T_d(\Gamma \to \infty) + (1/2)\ln(1 + 1/\Gamma) \qquad (6.15)$$

The second moment $<W^2>$ as a function of time is presented in Figure (6.5) for various values of Γ, namely, ∞, 10, 1, 0.1, and 0.01, which span very small to large correlation times. Further, $(\beta/k_B T)$ refers to the evolution of the normalized pseudo-thermodynamic energy level, and is decided by Equation (6.12a).

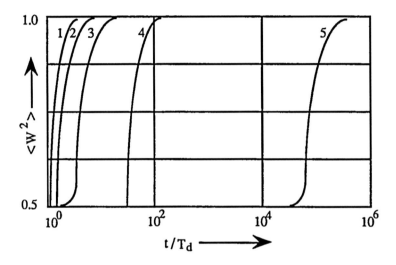

Figure 6.5 Evolution of the mean-squared value of W(t) for various discrete extents of correlation time (Γ)
(1. $\Gamma = 10^{-2}$; 2. $\Gamma = 10^{-1}$; 3. $\Gamma = 10^{0}$; 4. $\Gamma = 10^{+1}$; 5. $\Gamma = 10^{+4}$)

From Figure (6.5), it can be observed that the correlation time does not alter the qualitative aspects of the fluctuation behavior of the noise/disturbance. That is, in extensive terms, the onset of macroscopic order of neuronal state-transition is simply delayed when the correlation time increases.

6.5 Stochastical Instability in Neural Networks

Typically (artificial) neural networks are useful in solving a class of discrete optimization problems [34] in which the convergence of a system to a stable state is tracked *via* an energy function E, where the stable state presumably exists at the global minimum of E as mentioned in Chapter 4.

This internal state (in biological terms specified as the soma potential) of each neuron i is given by a time-dependent scalar value S_i; the equilibrium state is assumed as 0. The output of the cell (corresponding to spike or action potential frequency) σ_i is a continuous, bounded, monotonic function F. That is, $\sigma_i = F(S_i)$; and, in general, F is nonlinear. Thus, the output of the cell is a nonlinear function of the internal state.

Typically $F(x)$ is a sigmoid, taken conventionally in the hyperbolic tangent form as $(1/2) [1 + \tanh(\Lambda x)]$, or more justifiably as the Langevin function $L_q(\Lambda x)$ as described in Chapter 5. The coefficient Λ is the scaling factor which includes a pseudo-temperature corresponding to the Boltzmann (pseudo) energy of the system. Ideally, the rate of change of internal state is decided by the sum of the inputs from other neurons of the network in the form of weighted sum of firing rates by external sources (such as a constant bias) and by the inhibiting internal state:

$$(\partial S_i/\partial t) = \sum_j W_{ij}F(S_j) + \theta_i - S_i + \eta_i(t) \qquad (6.16)$$

where $\eta_i(t)$ represents the intracell disturbance/noise.

Upon integration (corresponding to a first-order low-pass transition with a time constant τ_0), Equation (6.16) reduces to:

$$S_i(t) = \int [\sum_j W_{ij}\sigma_j + \theta_i + \eta_i(t)]\exp[-(t - t')/\tau_0]dt'$$

$$(6.17)$$

With a symmetric weighting ($W_{ij} = W_{ji}$), Hopfield [31,36] defines an energy function (E) relevant to the above temporal model of a neuron-cell as:

$$E = -(1/2) \sum_i \sum_j W_{ij}\sigma_i\sigma_j - \sum_i (\theta_i + \eta_i) \sigma_i \qquad (6.18)$$

Convergence of the system to a stable state refers to E reaching its global minimum. This is feasible in the absence of the stochastical variable η_i (caused by the cellular disturbance/noise). However, the finiteness of η_i and the resulting strength of randomness could unstabilize the march of the network towards the global minimum in respect to any optimization search procedure. For example, a set of variables that can take only the two dichotomous limits (0 and 1) may represent possible solutions to discrete optimization problems. For each variable, a neuron can be assigned, with the optimization criteria specified by the energy function (E) of Equation (6.18). From this energy function, the coupling weights W_{ij} and the external input S_i can be decided or derived deterministically, in the absence of the disturbance/noise function η_i. That is, starting from an arbitrary initial state and with an appropriate scaling factor Λ assigned to the nonlinear function F, the neuron achieves a final stable state 0 or 1. Hence, a high output of a neuron (i, j), corresponding to an output close to its maximum value of 1, refers to an optimization problem similar to the considerations in assigning a closed tour for a traveling salesman over a set of N cities with the length of the tour minimized subject to the constraints that no city should be omitted or visited twice.

In the presence of $\eta(t)$, however, the aforesaid network modeling may become unstable; or the evolution of the energy-function decreasing monotonically and converging to a minimum would be jeopardized, as could be evinced from the following Hopfield energy functional analysis:

The evolution of E with progress of time in the presence of $\eta(t)$ having a dynamic state of Equation (6.16) and an input-output relation specified by $\sigma_i = F(S_i)$ can be written as:

$$E_t(t) = E(t) + \sum_i \int_{1/2}^{\sigma_i} F^{-1}(\sigma)\, d\sigma \qquad (6.19)$$

Using Equation (6.18),

$$\partial E_t(t)/\partial t = -\sum_i [(\partial S_i/\partial t)^2] F'(S_i) - \sum_i (\partial S_i/\partial t)\eta_i F'(S_i) - \partial \eta_i/\partial t F(S_i) \qquad (6.20)$$

The above equation permits $E_t(t)$ to decrease monotonically (that is, $E_t \leq 0$) and converge to a minimum only in the absence of $\eta(t)$. When such a convergence occurs for an increasing scaling factor of Λ (ultimately reaching infinity at the McCulloch-Pitts limit), $\int F^{-1}(\sigma)\, d\sigma$ would approach zero in any interval; and $(E - E_t)$ will therefore, become negligible for σ_i specified in that interval. That is, the minimum of E would remain close to that of E_t; but this separation would widen as the strength of η increases.

Failure to reach a global minimum in an optimization problem would suboptimize the solution search; and hence the corresponding computational time will increase considerably. Two methods of obviating the effect(s) of disturbances in the neural network have been suggested by Bulsara et al. [90]. With a certain critical value of the nonlinearity, the system can be forced into a double-well potential to find one or the other stable state. Alternatively, by careful choice of the input (constant) bias term θ_i, the system can be driven to find a global minimum more rapidly.

In the neural system discussed, it is imperative that the total energy of the system is at its minimum (Lyapunov's condition) if the variable W reaches a stable equilibrium value. However, the presence of $\eta(t)$ will offset this condition, and the corresponding wandering of W in the phase-plane can be traced by a phase trajectory. Such a (random) status of W and the resulting system instability is, therefore, specified implicitly by the joint event of instability pertaining to the firings of (presynaptic and/or postsynaptic) neurons due to the existence of synaptic connectivity.

Hence, in the presence of noise/disturbance, the random variate $W_{(1)}$ can be specified in terms of its value under noiseless conditions, namely, $W_{(2)}$ by a linear temporal gradient relation over the low-pass action as follows:

$$W_{(1)} = W_{(2)} + (\partial W_R/\partial t)t \tag{6.21}$$

where W_R is the root-mean-squared value of W, namely, $(<W^2>)^{1/2}$. By virtue of Equations (6.16-6.19, 6.21), the differential time derivative of the Lyapunov energy function E_t under noisy conditions can be written (under the assumption that the slope $\partial W_R/\partial t$ is invariant over the low-pass action) as:

$$
\begin{aligned}
\partial E_t(\sigma_i\sigma_j)/\partial t\big|_{noisy} = & [- \sum_i F'(S_i)\,(\partial S_i/\partial t)^2 - \sum_j F'(S_j)\,(\partial S_j/\partial t)^2] \\
& - 2(\partial W_R/\partial t)\big\{ [(\sum_i (\partial S_i/\partial t)F'(S_i)F(S_j) \\
& + \sum_j (\partial S_j/\partial t)F'(S_j)\,F(S_i)]t + F(S_i)F(S_j) \big\}
\end{aligned}
\tag{6.22}
$$

Hence, it is evident that as long as the temporal gradient of W_R in Equation (6.22) or the strength of noise η in Equation (6.20) is finite, the system will not reach *per se* the global minimum and hence the stable state.

6.6 Stochastical Bounds and Estimates of Neuronal Activity

Considering a neural network, the Hopfield energy surface as given by Equation (6.18) has the first term which refers to a *combinatoric* part whose minima correspond to solutions of a complex problem involving several interactive dichotomous variates. The second part of Equation (6.18) is *monadic* wherein such interactions are not present. This monadic term diminishes as the gain of the nonlinear process, namely, the scale factor $\Lambda \to \infty$ (as in the case of ideal McCulloch-Pitt's type of transitions). This also corresponds to Hopfield's pseudo-temperature being decreased in the simulated annealing process.

Suppose the variate W is uniformly distributed and the mean square deviation of W is designated as $M_{SW} = (<W> - W)^2$. The functional estimates of M_{SW} are bounded by upper and lower limits. For example, Yang et al. [91] have derived two possible lower bounds for M_{SW}, namely, the *Cramer-Rao* (CR) lower bound and the *information-theoretic* (IT) lower bound. Correspondingly, an asymptotic upper bound has also been deduced [91].

Further, considering a train of input sequences S_i stipulated at discrete time values t_i ($i = 1, 2, ..., N$), the weighting function W_i can be specified as a *linear least-squares estimator* as follows:

Let $\quad W_i = W_i{}^{NF} + (\partial W_R/\partial t)\, t_i + e_i \quad (1 \leq i \leq N)$

or $\quad W = at + e$ (6.23a)

where $W_i{}^{NF}$ is the initial intercept of W_i at $t_i = 0$ corresponding to the noise-free state (devoid of Fokker-Planck evolution) and e_i's are errors in the estimation; and:

$$a = \begin{bmatrix} W^{NF} \\ \partial W_R/\partial t \end{bmatrix} \tag{6.23b}$$

The minimum least-squares estimator of W_i, namely, W_e is therefore written as:

$$W_e = a_e t \tag{6.24a}$$

where $a_e = (H^T H)^{-1}(H^T W)$ with:

$$H = \begin{bmatrix} 1 & t_1 \\ 1 & t_2 \\ \vdots & \vdots \\ 1 & t_N \end{bmatrix} \text{ and } W = (W_1, W_2, ..., W_N)^T \tag{6.24b}$$

Hence explicitly:

$$H^T H = \begin{bmatrix} N & \Sigma t_i \\ \Sigma t_i & \Sigma t_i{}^2 \end{bmatrix} \text{ and } H^T W = \begin{bmatrix} \Sigma W_i \\ \Sigma t_i W_i \end{bmatrix} \tag{6.24c}$$

In ascertaining the best estimate of W_i, the slope $(\partial W_R/\partial t)$ should be known. For example, relevant to the data of Figure 6.5, the variation of W_R with respect to the normalized time function t/T_d (for different values of Γ) is depicted in Figure 6.6.

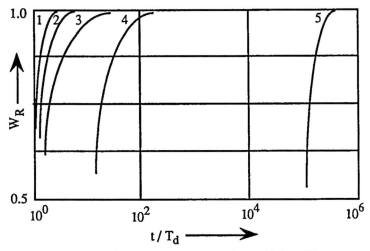

Figure 6.6 Evolution of root-mean-squared value W(t) for different extents
of the correlation time (Γ)
(1. $\Gamma = 10^{-2}$; 2. $\Gamma = 10^{-1}$; 3. $\Gamma = 10^{0}$; 4. $\Gamma = 10^{+1}$; 5. $\Gamma = 10^{+4}$)

As Γ increases, the corresponding time delay in neural response (T_d) decreases (or t/T_d increases) as in Figure 6.6. Hence, the functional relation between W_R and t/T_d can be "best fitted" as:

$$W_R(t) = \exp[+(T_{d\infty}/T_d + t/T_d]] \tag{6.25}$$

where $T_{d\infty}$ refers to the values of T_d as $\Gamma \to \infty$; and, $\exp(+T_{d\infty}/T_d)$ accounts for the constant of proportionality between W_R and $\exp(+t/T_d)$. Hence:

$$\partial W_R(t)/\partial t = +W_R(t)/T_d \tag{6.26}$$

In the presence of $\eta(t)$, Equation (6.17) can therefore be rewritten for the estimate of S_i, namely, S_{ei} as:

$$S_{ei}(t) = \int_0^{\tau_d} [\sum_j W_{eij}{}^{NF}\sigma_j + \theta_i] \exp[-(t-t')/\tau_0]dt'$$

$$+ \int_0^{\tau_d} \sum_j \sigma_j[W_R(t)/T_d](t-t') \exp[-(t-t')/\tau_0]dt' \tag{6.27}$$

where τ_d denotes the time of integration or the *low-pass action*. Further, the subscript e specifies explicitly that the relevant parameters are least-

123

squares estimates. Thus, the above equation refers to the time-dependent evolution of the stochastical variable S_i written in terms of its least squares estimate in the absence of $\eta(t)$ and modified by the noise-induced root-mean-squared value of W_R.

Upon integration, Equation (6.27) reduces at discrete-time instants to:

$$S_{ei}(t_i) = \tau_0 \sum_j (W_{eij}{}^{NF}(t_i)\sigma_{ej}(t_i) + \theta_i) \{ [\exp(-t_i/\tau_0)] [-1 + \exp(\tau_d/\tau_0)] \}$$

$$+ \sum_j \{ \sigma_{ej}(t_i) [W_R(t_i)/T_d(\Gamma)] \tau_0{}^2 \} \{ \sum_k^\infty (-1)^k [t_i - \tau_d/\tau_0]^{k+2}/[k!(k+2)] $$

$$- (-1)^k (t_i/\tau_0)^{k+2}/[k!(k+2)] \} \tag{6.28}$$

where $(t_i/\tau_0 \geq \tau_d/\tau_0)$.

Obviously, the first part of Equation (6.28) refers to noise-free deterministic values of S_i; and the second part is an approximated contribution due to the presence of the intracell disturbance/noise η. Implicitly, it is a function of the root-mean-squared value (W_R) of the stochastical variable W_i, spectral characteristics of η specified *via* the delay term $T_d(\Gamma)$, the time-constant of the low-pass action in the cell (τ_0), and the time of integration in the low-pass section (τ_d).

The relevant estimate of the Hopfield energy function of Equation (6.28) can be written as the corresponding Lyapunov function. Denoting the time-invariant constant term, namely, $\tau_0[\exp(\tau_d/\tau_0) - 1]$ as ϕ_1, the estimate of the Lyapunov function is given by:

$$E_{et}(t_i) = + (\phi_1/2) \sum_{i=1}^M \sum_{j=1}^N W_{eij}{}^{NF}(t_i)\sigma_{ej}(t_i)\sigma_{ei}(t_i)\exp(-t_i/\tau_0)$$

$$+ \phi_1 \sum_{i=1}^M [\exp(-t_i/\tau_0)]\sigma_{ei}(t_i) \theta_i + (\Lambda/2) \sum_{i=0}^M \int_0^{\sigma_{ie}(t)} F^{-1}(\mathcal{V}')d\mathcal{V}'$$

$$- (1/2) \sum_{i=1}^M \sum_{j=1}^N \phi_2 W_R(t_i)\sigma_{ei}(t_i)\sigma_{ej}(t_i) \tag{6.29a}$$

where $\tau_d \leq \tau_0$ and $\phi_2 [\tau_0{}^2/T_d(\Gamma)]$ represents the following expression:

$$\{ \sum_k^\infty (-1)^k [t_i - \tau_d/\tau_0]^{k+2}/[k!(k+2)] - (-1)^k (t_i/\tau_0)^{k+2}/[k!(k+2)] \}$$

$$\tag{6.29b}$$

In the absence of noise or disturbance (η), the energy function E as defined by Equation (6.18) (with the omission of η) has minima occuring at the corners of the N-dimensional hypercube defined by $0 \leq \sigma_i \leq 1$, provided θ_i's (i = 1, 2, ..., N) are forced to zero by a suitable change of coordinates.

In the event of the noise (η) being present, the estimated Lyapunov energy function given by Equation (6.29) cannot have such uniquely definable locations of minima at the corners. This is due to the presence of the noise-induced fourth term of the Equation (6.29a) correlating the ith and the jth terms, namely, $-(1/2) \Sigma^M_{i=1} \Sigma^N_{j=1} \phi_2 W_R(t_i)\sigma_{ei}(t_i)\sigma_{ej}(t_i)$, unless the second term involving θ_i and this fourth term are combined as one single effective bias parameter, $I_i(t_i)$; hence, I_i's can be set to zero *via* coordinate transformation forcing the minima to the corners of the N-dimensional hypercube as a first-order approximation discussed in the following section.

6.7 Stable States Search *via* Modified Bias Parameter

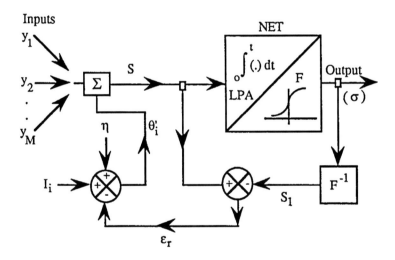

Figure 6.7 Linear recursive search of stable states

η: Noise; ε_r: Error; LPA: Low-pass action(integrator) F: Nonlinear estimator; I_i: Modified bias parameter [$I_i \rightarrow \theta_i$, as $\varepsilon_r \rightarrow 0$, $S_1 \rightarrow S$ and $\theta'_i \rightarrow \theta_i$ as $\eta \rightarrow 0$]

In the previous section, it is indicated that the presence of intracell disturbance η_i implicitly dictates the external bias parameter θ_i being modified to a new value specified as I_i. If the strength of randomness of the disturbance involved is small, an approximate (linear) recursive search for stable states is feasible. In general, the noise-perturbed vector σ when subjected to F^{-1} transformation yields the corresponding noise-perturbed value of S_1 as illustrated in Figure 6.7.

Hence, the summed input S and S_1 can be compared, and the corresponding error ε_r can be used to cancel the effect of the intracellular noise which tends to alter the value of the input bias θ_i to I_i as shown in Figure 6.7. The corresponding correction leads to $I_i \rightarrow \theta_i'$ $(\approx\theta_i)$. If necessary, a weighting W_I (such as linear-logarithmic weighting) can be incorporated on θ_i' for piecewise compatibility against the low to high strength of the randomness of the noise.

6.8 Noise-Induced Effects on Saturated Neural Population

The intracell disturbances could also affect implicitly the number of neurons attaining the saturation or the dichotomous values. Relevant considerations are addressed in this section pertaining to a simple input/output relation in a neuronal cell as depicted in Figure 6.8.

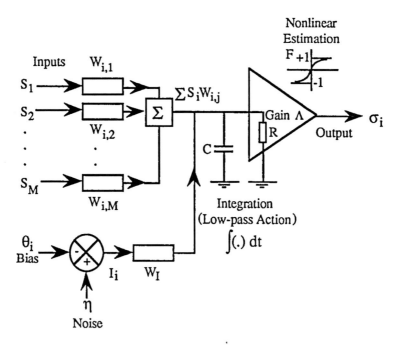

Figure 6.8 Network representation of neurocellular activity

The dynamics of neuron cellular activity can, in general, be written explicitly as [92]:

126

$$dS_i/dt + S_i/\tau_o = -1/\tau_o \sum_{\substack{j=1 \\ j \neq 1}}^{M} (\sigma_i + S_i)/W_{ij} + (I_i - S_i)/W_I$$

(6.30)

where τ_o is the time-constant (RC) of the integrator stated before and W_I is the weighting factor on the external bias θ_i (modified to value I_i due to the noise, η). The neuronal state change is governed between its dichotomous limits by a nonlinear amplification process (with a gain Λ) as follows:

$$\sigma_i = \begin{cases} \Lambda & \text{for } S_i > 1 \\ \Lambda(S_i) & \text{for } |(S_i)| \leq 1 \\ -\Lambda & \text{for } S_i < 1 \end{cases}$$

(6.31)

Over the transient regime of state change, the number of neurons attaining the saturation (or the dichotomous limits) would continuously change due to the nonlinear gain (Λ) of the system. Denoting the instants of such changes as a set of $\{t_k\}$, $k \in 0, 1, 2, ...$, at any instant t_k, the number of neurons still at subdichotomous limiting values is assumed as μ_k. Therefore, during the period $t_k \leq t \leq t_{k+1}$, the following state dynamics can be specified:

$$dS_i/dt = \begin{bmatrix} -(M+1)S_i - \sum_{j=1}^{M} \sigma_j + \Lambda(S_i) + I_i & \text{for } 1 \leq i \leq \mu_k \\ -(M+1)S_i - \sum_{j=1}^{M} \sigma_j + \chi_i & \text{for } \mu_k \leq i \leq M \end{bmatrix}$$

(6.32)

assuming that $W_{ij} = W_I$ and $\tau_o = 1$. Further, $\chi_i = I_i + \Lambda \, \text{Sign}(S_i)$ where:

$$\text{Sign}(S_i) = \begin{bmatrix} +1 & \text{if } S_i \geq 0 \\ -1 & \text{otherwise} \end{bmatrix}$$

(6.33)

The coupled relations of Equation (6.32) are not amenable for a single solution. However, as indicated by Yuan et al. [92], an intermediate function $U(t) = \Sigma^M_{j=1} \sigma_j$, with $|U(t)| \leq \Sigma^M_{j=1} |(\sigma_j)| \leq M\Lambda$ can be introduced in Equation (6.32) so as to modify it as follows:

$$dS_i/dt = \begin{bmatrix} (\Lambda - M - 1)S_i - U(t) + I_i & \text{for } 1 \leq i \leq \mu_k \\ -(M+1)S_i - U(t) + \chi_i & \text{for } \mu_k \leq i \leq M \end{bmatrix}$$

(6.34)

127

If $\Lambda > (M + 1)$ and $|S_i| \leq 1$, a relevant solution of Equation (6.34) indicates S_i growing exponentially. However, if $|S_i| > 1$, the dynamics of S_i become stable provided $I_i \rightarrow \theta_i$ with $\eta = 0$. At this stable state, considering the intermediate function $U(t) = \Lambda \Sigma^M_{i=1} \text{Sign}(S_i)$ and $|S_i| > 1$, for all $1 \leq i \leq M$, the dynamic solution of Equation (6.34) can be written as:

$$S_i = S_i(t_k)\exp[(\Lambda - M - 1)(t - t_k)]$$
$$+ [1/(\Lambda - M - 1)] \{(\exp[(\Lambda - M - 1)(t - t_k)] - 1\}$$
$$- U(t) u(t - t_k)\exp[(\Lambda - M - 1)t] \qquad \text{for } |S_i(t)| \leq 1, \; I_i \rightarrow \theta_i,$$

$$S_i = S_i(t_k) \exp[-(M + 1)(t - t_k)] + (\chi_i/(M + 1)) \{(1 - \exp\{-(M + 1)(t - t_k)\}$$
$$- U(t) u(t - t_k)\exp[-(M + 1)t] \qquad \text{for } |S_i(t)| > 1, \; I_i \rightarrow \theta_i,$$

$$(6.35)$$

As the network responds to an input vector S_i to yield a dichotomous vector σ_i, the initial condition set as $S_i(0)$ and the external bias parameter I_i ($\rightarrow \theta_i$) determine the division of neuronal states being "high" or "low". Yuan et al. [92] point out that the binary output vector σ has $M/2$ neuronal high states corresponding to the $M/2$ high state components of the bias input; and there are $M/2$ neuronal low states corresponding to the rest of the components of the bias input. In the event of θ_i being corrupted by an additive noise, the resulting input bias, namely, I_i, will upset this division of high and low level states in the output vector σ in a random manner which manifests as the neuronal instability.

6.9 Concluding Remarks

The inevitable presence of noise in a neural assembly permits the neurons to change their internal states in a random manner. The relevant state-transitional stochastic dynamics is governed by relaxational equation(s) of Langevin or Fokker-Planck types.

In general, the noise or intracell disturbances cited above could be gaussian, but need not be white. Such band-limited (colored) properties are intrinsic properties of the disturbances and are not influenced by the switching action of the state transition. Considering the colored noise situation, the Langevin and/or Fokker-Planck equation(s) can be solved by a scaling approximation technique.

The colored nature of the cellular noise also refers implicitly to the markovian nature of the temporal statistics of the action potentials which assume bistable values at random intervals. The weighting times in each state are exponentially distributed. Correspondingly, the onset of macroscopic order of neuronal state transition is simply delayed (in extensive terms) as the correlation time increases. The correlation time does not, however, alter the qualitative aspects of intracellular disturbances.

The effect of intracellular disturbances when addressed to artificial neural networks refers to stochastical instability in solving optimization problems. Such noise-induced effects would render the problem suboptimal with increased computational time.

Considering Hopfield networks, the presence of intracellular noise may not permit the network to settle at a global minimum of the energy function.

In terms of the Lyapunov condition, this nonrealization of a global minimum refers to the instability in the state transition process with specified lower and upper statistical bounds.

In the presence of noise, linear estimates of the input/output vectors of the neuronal network can be obtained *via* linear regression techniques. The corresponding estimate of the energy function indicates that the effect of intracellular disturbances can be implicitly dictated by modifying the constant (external) input bias to an extent proportional to the strength of the randomness.

The implications of this modified bias parameters are:

    a.   For small values of noise, a linear approximation of input-output relation leads to the feasibility of a recursive search for stable states *via* appropriate feedback techniques.

    b.   The modified bias parameter also alters the saturated neuronal state population randomly.

CHAPTER 7

Neural Field Theory:
Quasiparticle Dynamics and Wave Mechanics
Analogies of Neural Networks

7.1 Introduction

As a model, the neural topology includes a huge collection of almost identical interconnected cells, each of which is characterized by a short-term internal state of biochemical activity. The potential (or the state) at each cell is a dichotomous random variate; and, with a set of inputs at the synaptic junction pertaining to a cell, the state transition that takes place in the neuron progresses across the interconnected cells. Thus, the spatial progression of state-transitions represents a process of *collective movement*. Such spatiotemporal development of neuronal activity has been considered *via* partial differential equations depicting the diffusion and/or *flow* field considerations which refer to continuum theories (as opposed to detailed logic model of discrete neuronal units) and designated as *neurodynamics* or *neural field theory*. For example, as elaborated in Chapter 3, Beurle [42] proposed a flow model or wave propagation to represent the overall mean level of neuronal activity. Griffith [11-14] modeled the spatiotemporal propagation of neuroelectric activity in terms of an *excitation function* (Ψ_e) of the neurons and an *activity function* (F_a) concerning the soma of the neurons. He considered the neuronal spatial progression as an excitation that "is regarded as being carried by a continual shuttling between sources and field"; that is, the excitation (Ψ_e) creates the activity (F_a) and so on. He interrelated Ψ_e and F_a by $H_e\Psi_e = k_a F_a$ where H_e represents an "undefined" operator and developed a spatiotemporal differential equation to depict the neuronal flow. The efforts of Griffith were studied more elaborately by Okuda et al. [93] and the pursuant studies due to Wilson and Cowan [44] addressed similar spatiotemporal development in terms of a factor depicting the proportion of excitatory cells becoming active per unit time. Relevant equations correspond to those of *coupled van der Pohl oscillators*. Alternative continuum perspectives of viewing neuronal collective movement as the propagation of *informational waves* in terms of memory effects have also been projected in the subsequent studies [47].

In all the above considerations, the neural activity has been essentially regarded as a deterministic process with a traditional approach to neurodynamics based on dynamic system theory governed by a set of differential equations. However, the neural assembly in reality refers to a disorder system wherein the neural interactions closely correspond to the

130

stochastical considerations applied to interacting spins in the Ising spin-glass model(s) as discussed in Chapter 5. Such statistical mechanics attributions of the neuronal activity could warrant flow considerations analogous to particle dynamics of disorder systems. Hence, considered in the present chapter are *momentum flow* and *particle dynamics* analogies *vis-a-vis* the neuronal collective movement of state-transitional process in "bulk neural matter" viewed in continuum-based neural field theory.

7.2 "Momentum-Flow" Model of Neural Dynamics

As indicated by Peretto [38], the studies on the equilibrium and/or the dynamics of large systems (such as neural networks) require a set of *extensive quantities* to formalize the vectors depicting the equilibrium and/or dynamic state of the system. Such an extensive parameter embodies the distributed aspects of activity involving real (nonpoint/macroscopic) assembly. As well, it provides the relation between the localized phenomena (state instability, etc.) to the global picture of neural transmission, namely, the collective movement of neurons over the long-term memory space Ω representing the weights of the neuronal interconnections.

Discussed here is the possibility of representing the neural transmission or "propagation" as a collective progression of the state-transitions across the space Ω as analogous to a momentum flow so that an associated *wave function* formalism provides an alternative *extensive quantity* for the "input-outgo" reasoning pertaining to the neural assembly.

It is assumed that the neuronal aggregate is comprised of a large number M of elementary (cells) units i, $(i \in 1, ..., M)$; and the relevant dynamics of this neural system are viewed in the phase-space continuum Ω being dictated by x_i trajectorial variates with the associated momenta, p_i. The corresponding $H_N\{x, p\}$ refers to the Hamiltonian governing the equations of neural transmission. Considering analogously a *wave-packet* representation of the transport of M quasiparticles, the energy E associated with M units (on an extensive basis) of the neural assembly can be written as:

$$E = M\bar{h}\omega \qquad (7.1)$$

and the corresponding momentum is:

$$p = M\bar{h}k \qquad (7.2)$$

where ω is the "*angular frequency*", k is the "*propagation vector (constant)*" of the neuronal "wave" transmission, and $\bar{h}$ is a parameter (analogous to Planck's constant).

131

Then the associated rate of flow of energy (or power flow) can be specified by:

$$P = v_g M \bar{h} \omega \qquad (7.3)$$

where v_g is the "*group velocity*" of the wave analogously representing the neuronal flow; hence, the corresponding momentum flow can be expressed as:

$$T = v_g M \bar{h} k \qquad (7.4)$$

The input-output relation in a neuronal system refers essentially to a state-transition process depicting an energy E_1 (corresponding to a momentum p_1) changing to an energy E_2 with a momentum p_2, and the relevant conservation laws are therefore:

$$E_1 - E_2 = \bar{h} \omega \qquad (7.5a)$$
$$P_1 - P_2 = \bar{h} k \qquad (7.5b)$$

The above relations also represent analogously the neural transmission as a *quasiparticulate (or corpuscular) motion* in the domain Ω with a dual characterization as a wave as well. Hence the neuronal cell across which the state-transition between the dichotomous limits occurs, can be regarded as a *potential bidirectional well* wherein the "*neuronal particles*" reside.

The synaptic biochemical potential barrier energy Φ_{pB}, which represents a short-term activity in the neural complex, should be exceeded so that an output is realized as a result of the neuronal input energy, $E = \bar{h} \omega$. Correspondingly, the biochemical potential barrier energy can be depicted as $\Phi_{pB} = \bar{h} \omega_{pB}$. The output or no-output conditions (that is, the wave propagation or evanescent behavior), therefore, refers to $E \gtrless \Phi_{pB}$, respectively.

Now, denoting the neuronal momentum as $p = \bar{h} k$, the critical transition between a progressive and an evanescent (reflected) wave corresponds to the propagation vector $k \gtrless k_{pB}$, respectively.

Following the analysis due to Peretto [38], the set of internal states of a large neural network is designated by $S_i \{i \in 1, 2, ..., M\}$, which represents the internal state marker for the elementary unit i. Pertinent to the set $\{S_i\}$, an extensive quantity $Q(S_i)$ can be specified which is proportional to M, the size the of the system. The internal state has two dichotomous limits, namely, $+S_U$ and $-S_L$ associated with M_U and M_L

132

cellular elements (respectively) and $(M_U + M_L) = M$. Hence, in the bounded region Ω, (M_U/M) and (M_L/M) are fractions of neurons at the two dichotomous states, namely, S_U and S_L, respectively. It may be noted that in the deterministic model due to Wilson and Cowan [44], these fractions represent the proportions of excitatory cells becoming active (per unit time) and the corresponding inactive counterparts.

In terms of the phase space variable x and the associated momentum **p**, the probability distribution $\rho(\{x, p\}, t)$ refers to the probability of the systems being in the $\{x, p\}$ phase space, at time t. It is a localized condition and is decided explicitly by the stationary solution of the Boltzmann equation $(d\rho/dt = 0)$ leading to the *Gibbs distribution* given by:

$$\text{Limit}_{t \to \infty} \, \rho(\{x, p\}, t) = Z^{-1}\exp[-H_N(x, p)/k_B T] \qquad (7.6)$$

where Z refers to the partition function given by the normalization term $\Sigma_{\{x, p\}}\exp[-H_N(x, p)/k_B T])$; here, k_B is the pseudo-Boltzmann constant, T is the pseudo-temperature, and $H_N(x, p)$ refers to the Hamiltonian which is the single global function describing the dynamic system.

Pertinent to the kinetic picture of the neuronal transmission, the wave function Ψ and its conjugate Ψ^* are two independent variables associated with the collective movement of the neuronal process in a generalized coordinate system with **p** and x being the canonical momentum and positional coordinates, respectively. Hence, the following transformed equations can be written:

$$\Psi = (x - jp)/\sqrt{2} \qquad (7.7a)$$
$$\Psi^* = (x + jp)/\sqrt{2} \qquad (7.7b)$$

where **p** and x satisfy the classical commutation rule, namely, $[x, p] = 1$ and $[x, x] = [p, p] = 0$. Further, the corresponding Hamiltonian in the transformed coordinate system is:

$$U = \omega \, [p^2 + x^2]/2 \equiv \omega \, (\Psi\Psi^* + \Psi^*\Psi)/2 \qquad (7.8)$$

The canonical momentum **p** and the canonical coordinate x are related as follows:

$$dU/dp = dx/dt = \omega p, \quad dU/dx = -dp/dt = \omega x \qquad (7.9)$$

which are *Hamilton's first and second equations*, respectively; and the Hamiltonian **U** which refers to the energy density can be stated in terms of an amplitude function Φ as:

133

$$U = |\Phi|^2 \omega \tag{7.10}$$

The corresponding energy-momentum tensor for the neuronal transmission can be written as:

$$\begin{bmatrix} \tau & G \\ \mathscr{E} & U \end{bmatrix} \tag{7.11}$$

where

$$G = |\Phi|^2 k \qquad \text{Momentum density}$$
$$\mathscr{E} = |\Phi|^2 \omega \, v_g \qquad \text{Energy flux density}$$
$$\tau = |\Phi|^2 k \, v_g \qquad \text{Momentum flux density of the neuronal flow or flux.}$$

The above tensor is not *symmetric*; however, if the momentum density function is defined in terms of the weighting factor W as $G = |\Phi|^2 k/W^2$ (with the corresponding $\tau = |\Phi|^2 (k/W^2) v_g$, then the tensor is rendered symmetric.

The dynamics of the neuronal cellular system at the microscopic level can be described by the Hamiltonian $H_N(x_1, x_2, ..., x_M ; p_1, p_2, ..., p_M ; S_1, S_2, ..., S_M)$ with the state variables S_i's depicting the kinematic parameters imposed by the synaptic action. The link between the microscopic state of the cellular system and its macroscopic (extensive) behavior can be described by the partition function Z written in terms of Helmholtz free energy associated with the wave function. Hence:

$$Z(S_i, k_B T) = \int_{S_i}^{\infty} \exp\{-H_N(x, p; S_i)/k_B T\} dx dp \tag{7.12}$$

where $k_B T$ represents the (pseudo) Boltzmann energy of the neural system as stated earlier. On a discrete basis, the partition function simply represents the sum of the states, namely:

$$Z = \sum_i \exp(-E_i/k_B T) \tag{7.13}$$

where E_i depicts the *free energy* of the neuronal domain Ω_i. Essentially, Z refers to a controlling function which determines the average energy of the macroscopic neuronal system. There are two possible ways of relating the partition function *versus* the free energy adopted in practice in statistical mechanics. They are:

Helmholtz free energy:

$$E_H(S_i, k_B T) = -k_B T \ln[Z_{s_i}(S_i, k_B T)] \tag{7.14a}$$

Gibbs free energy:

$$E_G(f, k_B T) = -k_B T \ln[Z_f(f, k_B T)] \tag{7.14b}$$

The corresponding partition functions can be explicitly written as:

$$Z_{s_i}(S_i, k_B T) = \int_{\Omega_i} \exp[-H_{S_i}(x, p; S_i)/k_B T] dx dp \tag{7.15a}$$

and

$$Z_f(f, k_B T) = \int_{\Omega_i} \exp[-H_f(x, p; f)/k_B T] dx dp \tag{7.15b}$$

where f is the force vector.

The relevant Hamiltonians referred to above are related to each other by the relation,

$$H_f(x, p; f) \equiv H_{S_i} - S_i f \tag{7.16}$$

and the following Legendre transformations provide the functional relation between H_{S_i} *versus* the force vector f and S_i *versus* H_f.

$$f = \partial H_{S_i}(x, p; S_i) \tag{7.17a}$$

$$S_i = -\partial H_f(x, p; f)/\partial f \tag{7.17b}$$

Physically, for a given set of microscopic variables (x, p), the function H_{S_i} describes a system of neuronal "particles" with coordinates x_i's and momenta p_i's in interaction with the environment Ω_i having the states (S_i) stipulated by a set of kinematic parameters S_i's, $i \in 1, 2, ..., M$; and the Legendre transform $H_f(x, p; f)$ describes the same system in interaction with Ω_i with a dynamical force parameter f. Thus, H_{S_i} and H_f are alternative Hamiltonians to describe the neural dynamics associated with Ω_i.

7.3 Neural "Particle" Dynamics

The kinetic quasiparticle description (with a microscopic momentum-position attribution) of a neuronal phase space is *apropos* in depicting the corresponding localizable wave-packet, $\Psi(x, t)$. Considering the neuronal transmission across the ith cell similar to random particulate (*Brownian*) motion subjected to a quadratic potential, the *Langevin force equation* depicting the fluctuation of the state variable specifies that [87,94]:

$$S_i/S_o \equiv (\alpha m_N/k_B T)^{1/2}(v_i^* - v_i) \tag{7.18}$$

where S_o is a normalization constant, α is a constant dependent on the width of the potential barrier, m_N is the *pseudo-mass of the neuronal particle* and v_i^* is the critical velocity at which this "particle" in the absence of the effect of (thermal) random force will just reach the top of the barrier and come to rest. In the event of the neuronal energy E exceeding the barrier potential ϕ_{pB}, or $v_i > v_i^*$, the corresponding transmission function C_{Tni} is given by [87,94]:

$$C_{Tni} = (1/2)\{1+\text{erf}[(\alpha m_N/k_B T)^{1/2}(v^* - v_i)]\} \tag{7.19}$$

The transmission function indicated above specifies implicitly the nonlinear transition process between the input-to-output across the neuronal cell. The motion of a "neuronal particle" can also be described by a wave function $\Psi_i(x, t) = \exp(-jE_i t/\bar{h})\Psi_i(x)$. The eigenfunction $\Psi_i(x)$ is a solution to:

$$(-\bar{h}^2/2m_N)[d^2\Psi_i(x)/dx^2] = E\Psi_i(x) \qquad x < 0 \tag{7.20a}$$

and

$$(-\bar{h}^2/2m_N)[d^2\Psi_i(x)/dx^2] = (E_i - \phi_{pB})\Psi_i(x) \qquad x > 0 \tag{7.20b}$$

assuming that at $x = 0$ crossing of the potential barrier occurs depicting a neuronal state transition.

The traveling wave solution of Equation (7.20a) in general form is given by:

$$\Psi_1(x) = A \exp(jp_1 x/\bar{h}) + C_{Rn} \exp(-jp_1 x/\bar{h}) \qquad x < 0 \tag{7.21a}$$

136

where $p_1/\bar{h}$ is the momentum equal to $(2m_N E_1)^{1/2}/\bar{h}$; and C_{Rn} is the reflection coefficient.

Similarly, for $x > 0$:

$$\Psi_2(x) = (1 - C_{Rn})\exp(+jp_2x/\bar{h}) \tag{7.21b}$$

where $p_2/\bar{h} = [2m_N(E_2 - \phi_{pB})]^{1/2}/\bar{h}$; and $(1 - C_{Rn})$ is the transmission coefficient.

Classically, the neural state transmission corresponding to the "neuronal particle" entering the (output) region $x > 0$ has a probability of one. Because of the presumed wave-like properties of the particle, there is a certain probability that the particle may be reflected at the point $x = 0$ where there is a discontinuous change in the regime of (pseudo) de Broglie wavelength. That is, the probabililty flux incident upon the potential discontinuity can be split into a transmitted flux and a reflected flux. When $E \approx \phi_{pB}$, the probability of reflection would approach unity; and, in the case of $E >> \phi_{pB}$, it can be shown that [95]:

$$C_{Rn} = \left\{[1 - (1 - \phi_{pB}/E)^{1/2}]/[1 + (1 - \phi_{pB}/E)^{1/2}]\right\}^2 \tag{7.22}$$

or even in the limit of large energies (or $E >> \phi_{pB}$), the (pseudo) de Broglie wavelength is so very short that any physically realizable potential ϕ changes by a negligible amount over a wavelength. Hence, there is total potential transmission with no reflected wave corresponding to the classical limit. Further, the transmission factor (for $x > 0$) can be decided by a function Func(.) whose argument $\mathcal{B}$ can be set equal to:

$$\mathcal{B} = (2/3) [(2m_N/\bar{h}) (E/\phi_{pB} - 1)]^{1/2}(a_{pB})^{3/2} \tag{7.23}$$

where a_{pB} refers to the *width of the potential barrier*.

Therefore, the transmission factor of Equation (7.19) can be rewritten in terms of the energy, mass and wave-like representation of neuronal transmission as:

$$C_{Tn} = (1/2)\left\{1 + \text{erf}\left(2/3[(2m_N a_{pB}^2/\bar{h}^2)\phi_{pB}]^{1/2} [E/\phi_{pB} - 1]^{1/2}\right)\right\} \tag{7.24}$$

In terms of neuronal network considerations, C_{Tn} can be regarded as the time-average history (or state-transitional process) of the activation induced updates on the state-vectors (S_i) leading to an output set, σ_i. This average is decided explicitly by the modified Langevin function as indicated in Chapter 5. That is, by analogy with particle dynamics wherein the collective response is attributed to nonlinear dependence of forces on positions of the particles, the corresponding statistics due to Maxwell-Boltzmann could be extended to neuronal response to describe the stochastic aspects of the neuronal state-vector. The ensemble average which depicts the time-average history thereof (regarding the activation-induced updates on state-vectors) is the modified Langevin function, given by [16]:

$$L_q(\beta_G S_i/k_B T) = \left\{ (1 + 1/2q)\coth[(1 + 1/2q)\beta_G S_i/k_B T] \right.$$
$$\left. - (1/2q)\coth([\beta_G S_i/2qk_B T]) \right\} \qquad (7.25)$$

where β_G is a scaling factor and $(\beta_G/k_B T)$ is a nonlinear (dimensionless) gain factor Λ. This modified Langevin function depicts the stochastically justifiable squashing process involving the nonlinear (sigmoidal) gain of the neuron unit as indicated in Chapter 5. Further, the modified Langevin function has the slope of $(1/3 + 1/3q)$ at the origin which can be considered as the order parameter of the system. Therefore, at a specified gain-factor, any other sigmoidal function adopted to depict the nonlinear neuronal response should have its argument multiplied by a factor $(1/3 + 1/3q)$ of the corresponding argument of the Langevin function.

Heuristically, the modified Langevin function denotes the transmission factor across the neuronal cell. Hence, writing in the same format as Equation (7.19), the transmission factor can be written in terms of the modified Langevin function as follows:

$$C_{Tn} = 1/2 \left[(1 + L_q(\beta_G S_i/k_B T)] \right. \qquad (7.26a)$$

Comparing the arguments of Equations (7.24) and (7.26a), it can be noted that:

$$\Lambda_{S_i} = (\beta_G S_i/k_B T) \equiv 2 [2m_N a_{pB}^2/\bar{h}^2 (E_i - \phi_{pB})]^{1/2} \qquad (7.26b)$$

omitting the order parameter which is a coefficient in the argument of Equation (7.24), due to the reasons indicated above. Hence, it follows that:

$$\Lambda = 2 (2m_N a_{pB}^2 \phi_{pB}/\bar{h}^2)^{1/2} \text{and } S_i = (E_i/\phi_{pB} - 1)^{1/2} \qquad (7.26c)$$

Thus, in the nonlinear state-transition process associated with the neuron, the limiting case of the gain $\Lambda \to \infty$ (McCulloch-Pitts regime) corresponds to the potential barrier level $\phi_{pB} \to \infty$. This refers to the classical limit of the (pseudo) de Broglie wavelength approaching zero; and each neuronal state S_i is equitable to an energy level of E_i. That is, for the set of neuronal state-vectors $\{S_i\} \Leftrightarrow \{E_i\}$.

The McCulloch-Pitts depiction of a neuron (best known as the "formal" or "mathematical" neuron) is a logical and idealized representation of neuronal activity. It purports the idealization of a real neuron with the features of being able to be excited by its inputs and of giving an output when a threshold is exceeded. This all-or-none response, although it provides a digital machine logic mathematically with ease of tracking the state-transitions in the neuronal transmission, it is rather unrealistic in relation to its time dependence. That is, the McCulloch-Pitts model presents results on the state-transitional relaxation times which are astronomically large in value, "which is not obvious" in the real neuron situation as also observed by Griffith [11,12]. In the present modeling strategies, the nonrealistic aspects of the McCulloch-Pitts model *vis-a-vis* a real neuron is seen due to the fact that only in the limiting case of the (pseudo) de Broglie wavelength approaching zero, the nonlinear gain (Λ) of the state-transition process would approach infinity (corresponding to the McCulloch-Pitts regime). However, this is only possible when the cellular potential barrier value ϕ_{pB} approaches infinity which is rather physically not plausible. As long as ϕ_{pB} has a finite value less than E_i, Λ is finite as per the above analysis confirming a realistic model for the neuronal activity rather than being the McCulloch-Pitts version. In terms of magnetic spins, the spontaneous state transition (McCulloch-Pitts' model) corresponds to the thermodynamic limit of magnetization for an infinite system (at temperature below the critical value). This infinite system concept translated into the equivalent neural network considerations, refers to the nonlinear gain (Λ) of the network appproaching infinity.

The foregoing deliberations lead to the following inferences:

- The dynamic state of neurons can be described by a set of extensive quantities *vis-a-vis* momentum flow analogously equitable to a to quasiparticle dynamics model of the neuronal transmission, with appropriate Hamiltonian perspectives.

- Accordingly, the neural transmission through a large interconnected set of cells which assume randomly, a dichotomous short-term state of biochemical activity can be depicted by a wave function representing the particle motion.

- Hence, considering the wave functional aspects of neuronal transmission, corresponding eigen-energies can be stipulated.

- Further, in terms of the Hamiltonian representation, of neural dynamics, there are corresponding free-energy (Helmholtz and Gibbs' versions) and partition functions.

- Relevant to quasiparticle dynamics representation, the neural transmission when modeled as a random particulate (Brownian) motion subjected to a potential barrier at the neuronal cell, a Langevin force equation can be specified for the state-transition variable in terms of a "neuronal mass" parameter; and, the transmission (excitatory response) across the cell or nontransmission (inhibitory response) is stipulated by the neuronal particle (dually regarded also as a wave with an eigen-energy) traversing through the cellular potential or reflected by it. The quasiparticulate and wave-like representation of neuronal transmission leads to the explicit determination of transmission and reflection coefficients.

- On the basis of particulate and/or wave-like representation of neuronal transmission and using the modified Langevin function description of the nonlinear state-transition process, a corresponding gain function can be deduced in terms of the neuronal mass and the barrier energy. Relevant formalism specifies that the gain function approaching infinity (McCulloch-Pitts' regime) corresponds to the potential barrier level (ϕ_{pB}) at the cell becoming infinitely large in comparison with the eigen-energy (E_i) of the input (which is not however attainable physically).

- In terms of (pseudo) de Broglie's concept extended to the dual nature of neurons, the spontaneous transition of McCulloch-Pitts' regime refers to the wavelength becoming so very short (with $\phi_{pB} \ll E_i$) that any physically realizable potential would change only by a negligible amount over a wavelength. This corresponds to the classical limit of transmission involving no reflection.

- The "size" of the wave-packet associated with neuronal quasi-particle transmission can be related to the range of the interaction and the momentum of the incident particle. When the spread is minimum, the passage of a wave packet can be considered

140

as the neuronal transmission with a state-transition having taken place. This can be heuristically proved as follows.

The extent of the cluster of cells participating in the neuronal transmission refers to the average range of interaction (equal to $<x_{ij}>$); and the corresponding size of the wave-packet emulating the neuronal transmission can be decided as follows: Suppose $(\Delta x_{ij})^{in}$ is the spatial spread of the incident wave-packet at the site, i. After the passage through the cell (local interaction site), the corresponding outgoing wave-packet has the spread given by [3]: $[(\Delta x_{ij})^{out}]^2 \approx [(\Delta x_{ij})^{in}]^2 + (\Delta v_{ij})^2(\tau_{ij})$ where τ_{ij} is the ith to j^{th} site-transit time; and the velocity term Δv_{ij} is dictated by the *uncertainity principle*, namely, $\Delta v_{ij} \approx \hbar/m_N \Delta x_{ij}$. Also $\tau_{ij} \approx$ $<x_{ij}>/v_{ij} = <x_{ij}> m_N/p$. Hence $(\Delta x_{ij})^{out}$ has a minimum when $(\Delta x_{ij})^{in} = (<x_{ij}> \hbar/p)^{1/2}$. The corresponding momentum spread is, therefore, $\Delta p_{ij} \approx (p\hbar /<x_{ij}>)^{1/2}$.

The aforesaid parameter spreading implicitly refers to the *smearing parameter* defined by Shaw and Vasudevan [76] in respect to Little's analogy [33] of neurons and and Ising glass spins. The *ad hoc* parameter $\beta = 1/k_B T$ in Little's model represents the fluctuation governing the total (summed up) potential gathered by the neuron in a time step taken in the progression of neuronal activity. Shaw and Vasudevan relate the factor $1/k_B T$ to the gaussian statistics of the action potential and the poissonian process governing the occurrence of the rate of chemical quanta reaching the postsynaptic memory and electrically inducing the postsynaptic potential. The relevant statistics elucidated in [76] referred to the variations in size and the probability of release of these quanta manifesting as fluctuations in the postsynaptic potentials (as observed in experimental studies).

It can be observed that this smearing action is seen implicitly in terms of wave function attributes specified as the spreads $(\Delta x, \Delta p)$. Thus $\Lambda = \beta_G \beta = \beta_G/k_B T$ is controlled by the smearing action. That is, in the total absence of fluctuations in the postsynaptic potentials, $k_B T \rightarrow 0$; and $\Lambda \rightarrow \infty$ depicts McCulloch-Pitts' regime. In the real neurons, the inevitable fluctuations in the summed up potential would always lead to a finite smearing with corresponding finite spreads in the wave function parameters leading to nonspontaneous state-transitions with a finite value Λ.

7.4 Wave Mechanics Representation of Neural Activity

As discussed earlier, the biological neural assembly refers to a complex linkage of an extremely large number of cells having synaptic junctions across which the neuronal transmission takes place. Such a transmission is governed by a complex biochemical (metabolic) activity and manifests as a train of spiky impulses (action potentials). That is, the operation of neural networks, in essence, refers to the all-or-none response of the basic units (neurons) to incoming stimuli. A neuron is activated by the flow of chemicals across synaptic junctions from the axons leading from other neurons. These electrical effects which reach a neuron may be excitatory or inhibitory postsynaptic potentials. If the potential gathered from all the synaptic connections exceeds a threshold value, the neuron fires and an action potential propagates down its one output axon which communicates with other neurons *via* tributary synaptic connections. The neuronal topology, as a model, includes therefore a huge collection of almost identical (cell) elements; and each such element is characterized by an internal state (of activity) with transitional dynamics and connected to each other *via* synaptic interactions.

In general, the interconnected set of neurons (Figure7.1) represents a disordered system with the following attributes: (1) In the assembly of M neurons, the ith neuron is characterized by the ordered pair (S_i, W_{ij}) representing the short-term (memory) state and the long-term (memory) state respectively; (2) The short-term (memory) state S_i refers to the state of the neuron's local (biochemical) activity; (3) The long-term (memory) state W_{ij} is a multidimensional vector modeling the synaptic weights associated between the ith and the other interconnected neurons; (4) With a feedback that exists between the output to the input, the neural assembly represents a recurrent network wherein the input is updated or modified dynamically by the feedback from the output. That is, in response to an applied input vector, each neuron delivers an output; and the ith neuron's output is taken commonly as a sigmoid function, $F(S_i)$, of the short-term state. The state equation for a neuron can be written as:

$$\partial S_i/\partial t = \sum_j W_{ij}F(S_j) - S_i + \theta_i \qquad (7.27)$$

where θ_i is a constant external bias; and (5) the neuronal inputs represented by a set of vectors $\{y_i^{in}\}$ are a sequence of statistically independent entities which are also distribution-ergodic.

The activity of a biological neuron as summarized above in an organization of a large neuronal assembly represents a cooperative process in which each neuron interacts with many of its neighbors; and, as dictated by a threshold, the input-output response of a neuron was represented as a simple dichotomous state system — active and inactive by McCulloch and Pitts. This "formal" or "logical" neuron model has been developed into a more comprehensive model since then with the state equation given by Equation (7.27); and the corresponding Hamiltonian associated with the neuronal interaction is written as:

$$H_N = - \sum_i \sum_j W_{ij} S_i S_j - \sum_i \theta_i S_i \qquad (7.28)$$

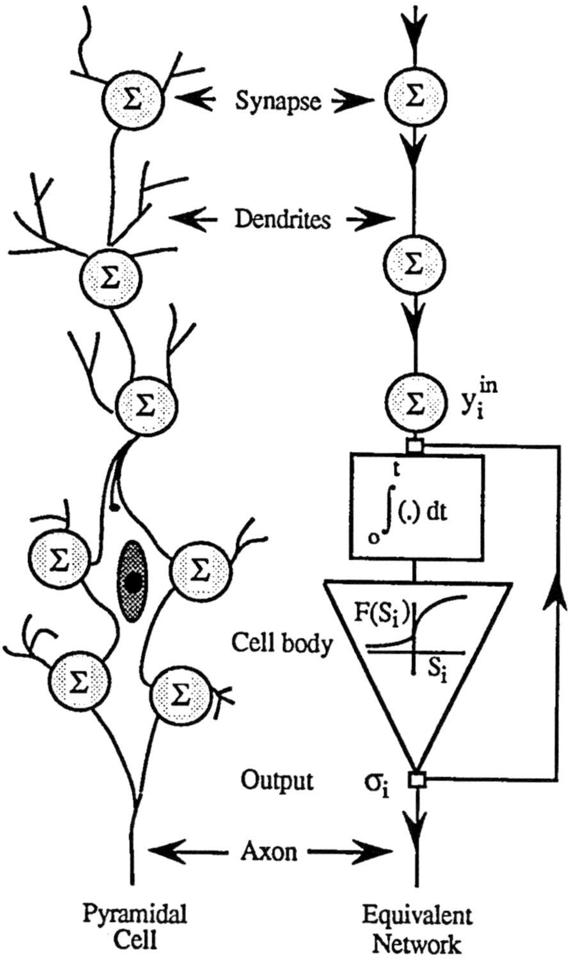

Figure 7.1 Pyramidal interconnection of neural cells and its network representation

In many respects, the aforesaid mathematical depiction of the neuron is analogous to the magnetic spin model in which a large ensemble of magnetic dipoles/domains interact with each other in varying weights as detailed in Chapter 5. In addition to the statistical mechanics characterization of neuronal activity, the input-output relation depicting the collective response of the neuronal aggregates has also been treated as due to a flow of flux spreading across the interacting units as described earlier in this chapter.

In view of these considerations which display the analogies between statistical mechanics and neuronal dynamics indicating the feasibility of modeling the neuronal activity by a "flow system", an alternative approach based on wave mechanical perspectives in conjunction with the associated wave function(s) depicting the neural transmission can be considered as discussed in the following sections.

7.5 Characteristics of Neuronal Wave Function

The neural activity as modeled here refers to the motion of the neuronal wave or the collective movement of state-transitions through a set of interconnected sites (neuronal cells) in the domain Ω. These cells are at random dichotomous potentials as a result of short-term biochemical ionic activity across the cellular membrane. Each time when the neuronal assembly is stimulated by an input, it corresponds to a set of inputs causing a "collective movement" of neurons through a domain which can be designated as the long-term (memory) space (Ω). This weight-space, in which both the moving state vector $\{W_{ij}\}$ and the applied input vector exist (Figure 7.2), constitutes a continuum wherein the evolution of neuronal distribution and a dynamic activity persist.

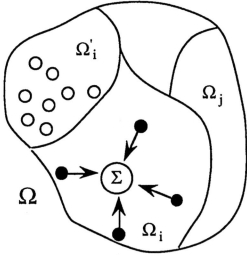

O   Inactive Neurons
●   Active Neuron Set $\{W\}$
Σ Input Vector Set $\{y\}$

Figure 7.2 Long-term neural activity space (Ω)

144

The proliferation of neuronal wave through this system can be described by a time-dependent wave function $\Psi(x, t)$ namely:

$$\partial\Psi(x, t)/\partial t = - \left(\tau_N v^2\right)\nabla^2\Psi(x, t) - \left(1/\tau_N\right)E(x, t)\Psi(x, t)$$

$$(7.29)$$

where $E(x, t)$ refers to the random site potential with a given statistics, v is the (pseudo) velocity of neuronal wave propagation, and $(1/\tau_N)$ is the rate at which the neuronal transmission is set inhibitory at the sites (cells) wherein the random potential $E(x)$ assumes a zero value with a probability w; and, at the sites where $E(x)$ exceeds a threshold barrier potential ϕ_{pB}, there is an excitatory process permitting the neuronal transmission with a probability $(1 - w)$. The random potential at a cellular site, namely, $E(x)$, is assumed to be uncorrelated at different sites and dichotomous in nature; that is, it takes values zero or ϕ_{pB} (with probabilities w and $(1 - w)$ respectively). The integral $\int\Psi(x, t)dx$ represents the total neuronal transmission probability.

Assuming the initial condition for $\Psi(x, t = 0)$ as independent of the site coordinate x, then at an ith site with a random potential $E_i(x)$, the solution for $\Psi_i(x, t)$ can be written explicitly in terms of a complete set of eigenfunctions and eigenvalues, namely:

$$\Psi_i(x, t) = \sum_{m=0}^{\infty} C_m^i \exp(-\Phi_m^i t)\, \Psi_m^i(x)$$

$$(7.30)$$

where $C_m^i = \int \Psi_m^i(x)dx$; and the eigenvalues Φ_m^i are all positive. Further, the asymptotic expansion of $\Psi_i(x, t)$ is decided predominantly by the lowest eigenvalues. As defined earlier, the total transmission probability for a given site is $\Psi_i(t) = \int\Psi_i(x, t)dx$. The corresponding ensemble-average behavior of the neuronal transmission over the entire neuronal cell aggregates can be specified by the first cumulant $[K_2(t) - K_1^2(t)]$, where $K_1(t)$ and $K_2(t)$ are the first and second moments of $\Psi_i(x, t)$ respectively; and this first cumulant refers to the lower bound of the neuronal transmission statistics. It is much larger than the mean value, $K_1(t)$. That is, the fluctuations of neuronal transmission dominate the mean level of transmission. In other words, the random neuronal transmission across the interconnected cells over the long-term (memory) space Ω or the collective movement of state-transitions is a zero-mean stochastical process. Correspondingly, the wave function set forth by the random potentials at the cellular sites has a *non-self-averaging* behavior.

145

The incident (or incoming) collective movement of neurons in the domain Ω has its energy relation specified again in terms of a wave function Ψ similar to the conventional Schrödinger wave equation:

$$\nabla^2\Psi + (1/\bar{h}^2 v^2)(E^2 - \Phi_{pB}^2)\Psi = 0 \qquad (7.31)$$

where v, as indicated earlier, refers to the velocity of the "neuronal wave"; and the "connectivity" of neuronal transmission between the cells, therefore, corresponds to a scalar weighting factor (similar to the refractive index of a medium) given by $W = [1 - (\Phi_{pB}/E)^2]^{1/2}$.

Considering the analogy with the Ising spin model, the interconnected set of a large number of M neurons (with the state transition at a localized site, i) has a Hamiltonian or the extensive quantity given by [38]:

$$H_N(S_i) = (-1/M)\sum_i\sum_j W_{ij}S_iS_j - \sum_i\theta_iS_i \qquad (7.32)$$

where W_{ij} is the weighting factor, and $\Sigma\theta_i\sigma_i$ is the Hamiltonian introduced by an external field (bias) θ_i at site i. In terms of the extensive parameter (or the Hamiltonian) H_N, the eigenvalue equation can be written as: $H_N\Psi_i = E_i\Psi_i$, where i indexing refers to the set of eigenstates.

Considering the neuronal wave equation (Equation 7.29), the progressive (traveling) wave solution given by Equation (7.30) can be explicitly written as:

$$\Psi(x, t) = \Phi(x, t)\exp\{j(\omega t - k{\cdot}x + \theta_a)\} \qquad (7.33)$$

where the propagation vector (constant) k is decided by the boundary conditions relevant to the wave function Ψ. Further, the quantity θ_a refers to the phase constant associated with the amplitude of the wave, and Φ refers to the amplitude function of the wave.

The selective wave functions Φ_i (or mode functions), which are the eigen-solutions of the wave equation and propagate across the domain Ω, correspond to the excitatory neuronal transmission. Those which are cut off can be considered as having faced an inhibitory process and represent the evanescent waves. Over the subregions of Ω ($\bigcup_{i=1}^{M}\Omega_i = \Omega$) with $\Omega_i \cap \Omega_j = 0$ for $i \neq j$, and M as indicated earlier, refers to the total number of subregions or cells (Figure 7.2). For the ith subregion, field Ψ_i is defined as $\Psi_i = \Psi(\Omega_i)$ or zero for outside Ω_i; and $\Psi = \Sigma_i^M \Psi_i$ with the summation representing the geometrical conjunction or the synaptic junction wherein the superposition of neuronal inputs occurs.

Each Φ_i being orthonormal to all other mode functions permits the incident (incoming) wave function Ψ_{in} to be expanded more explicitly in reference to the two domains Ω_i ($i \in M_U$) and Ω_i' ($i' \in M_L$) *via* spatial harmonics as:

$$\Psi_{in} = \sum_i^{M_U} \sum_n A_{ni} \Phi_{ni} + \sum_{i'}^{M_L} \sum_v B_{vi'} \Phi_{vi'} \tag{7.34}$$

where ($n = 0, 1, ..., \infty$), ($v = 0, 1, ..., \infty$) and A_n and B_v are the amplitudes of the relevant modes.

The $\{(\Psi_i)\}$ components in the ith subregion correspond to the excitatory process leading to the final outcome as ($+S_U$); and ($\Psi_{i'}$)'s are the reflected waves which are evanescent. They do not contribute to the output, and the relevant inhibitory process renders the dichotomous state as ($-S_L$). The existence of (Ψ_i)'s and ($\Psi_{i'}$)'s is dictated by the relevant continuity conditions. In terms of energy density functions, the conservation relation can be written in terms of the modal functions as:

$$\int_\Omega |\Phi_{in}|^2 k \, v^2 \, d\Omega - \sum_{n=1}^\infty \int_{\Omega_i} |\Sigma\Phi_i|^2 k_n v_n^2 d\Omega$$

$$\equiv \sum_{n=1}^\infty \int_{\Omega_i'} |(\Sigma\Phi_{i'})|^2 k_n v_n^2 d\Omega \tag{7.35}$$

If the amplitude of the incident wave function is set equal to $a(\Omega_i/\Omega)$ where (Ω_i/Ω) is the fraction depicting the spatial extent (or size) of the ith neuron in the state-space assembly of Ω, then $\Sigma\Phi_i$ and $\Sigma\Phi_{i'}$ can be proportionately set equal to $C_{Tn} a(\Omega_i/\Omega)$ and $C_{Rn} a(\Omega_i/\Omega)$, respectively. Here, C_{Tn} and C_{Rn} as mentioned earlier are coefficients of transmission and reflection, respectively, so that $C_{Tn} = (1 - C_{Rn})$.

7.6 Concepts of Wave Mechanics *versus* Neural Dynamics

The concepts of wave mechanics described above can be translated into considerations relevant to the neuronal assembly and/or a large (artificial) neural network. Considering the neuronal dynamics, the kinetics of the neuronal state denoted by a variable S_i and undergoing a transition from S_1 to S_2 can be specified by a transition probability (or the probability per unit time that a transition would take place from state S_1 to state S_2), namely, $w_i(S_1, S_2)$; and, in terms of wave mechanics perspective, w_i is given by the *Fermi golden* rule, namely:

147

$$w_i(S_1, S_2) = (2\pi/\bar{h})[\Phi_i(S_1, S_2)]^2 \delta(E_{S1} - E_{S2} \pm \bar{h}\omega)$$

$$(7.36)$$

where Φ_i is the eigen-potential function specified by Equation (7.35), E_{S1} and E_{S2} are energy level transitions, and the delta function guarantees the energy conservation.

In terms of this transitional probability, the rate of change of the probability distribution function ρ (at a given time, t) of the neuronal state (corresponding to the subregion Ω_i) can be written via the well-known Boltzmann equation, namely:

$$d\rho(S_1, t)/dt = \int \{w(S_2, S_1)\rho(S_2, t)[1 - \rho(S_1, t)]\}dS_2$$
$$- \int \{w(S_1, S_2)\rho(S_1, t)[1 - \rho(S_2, t)]\}dS_2 \quad (7.37)$$

The above Equation (7.37) specifies the net change in ρ at the same instant (markovian attribution) as the excitatory process permits the progressive neural transmission and the inhibitory process sets an evanescent condition of inhibiting the neural transmission.

Under the equilibrium condition, setting $d\rho/dt = 0$ yields:

$$w(S_2, S_1)\rho^o(S_2)[1 - \rho^o(S_1)] = w(S_1, S_2)\rho^o(S_1)[1 - \rho^o(S_2)]$$

$$(7.38)$$

where the superscript o refers to equilibrium status. Assuming that the equilibrium values $\rho^o(S_1)$ and $\rho^o(S_2)$ are decided by Boltzmann distribution, $\rho^o(S_{1,2}) = 1/\{1 + \exp[-(E_{S1,2} - \phi_{pB})/k_BT]\}$, it follows that $w(S_2, S_1)$ $\exp[(E_{S1} - E_{S2})/k_BT] = w(S_1, S_2)$; and this relation (known as the *principle of detailed balance*) must be satisfied regardless of the microscopic origin of neuronal interactions, as has been observed by Peretto [38] . Further, if the energy of the equilibrium state (S_0) is much larger than the other two dichotomous states S_1 and S_2, $w(S_2, S_0) << w(S_0, S_2)$, the solution of master equation (Equation 7.37) leads to: $\rho(S_0, t) = \rho(S_0, 0)$ $\exp\{-t/\tau_{S0}\}$. That is, the probability of neurons being in the state $|S_0>$ will decay exponentially with a time-constant τ_{S0}; or from the neural network point of view it refers implicitly to the integration process (*low-pass action*) associated with neuronal input-output relation.

In terms of the macroscopic potential function $\Phi(S_0)$ in the sample domain Ω_i, specified by $\Phi(S_0) = \Sigma_{S_i}\Phi(S_i)\rho_{s_o}(E_{S_i})$, the time dependency of the neuronal transition process can also be written as:

$$\Phi(S_i, t) = \Phi(S_0) + [\Phi(S_i, t = 0) - \Phi(S_0)]\exp[-t/<\tau>]$$

(7.39)

where $<\tau>$ is the average (energy) relaxation time or the time of integration involved in neural transmission.

In a neuronal aggregate of M cells with the dichotomous state of activity, there are 2^M possible different states which could be identified by $S = 1, ..., 2^M$ associated with an M-dimensional hypercube. A superficial analogy here with the quantum statistical mechanics situation corresponds to a set of M subsystems, each having two possible quantum states as, for example, a set of M atoms each having a spin 1/2 [13,32].

Each of the 2^M states has a definite successor in time so that the progress of the state-transition process (or the neuronal wave motion) can be considered as a sequence $i_2 = z(i_1) \rightarrow i_3 = z(i_2) \rightarrow ...$, and so on. Regarding this sequence, Griffith [13] observes that in the terminal cycle of the state-transitional process, there are three possible situations, namely, a state of equilibrium with a probability distribution $\rho(S_0)$; and the other two are dichotomous states, identified as $S_1 \Rightarrow +S_U$ and $S_2 \Rightarrow -S_L$ with the statistics $\rho(S_1)$ and $\rho(S_2)$, respectively.

In computing the number of states which end up close to the equilibrium at the terminal cycle, Griffith [13] observes the following fundamental difference between the neural and the quantum situation. From the quantum mechanics point of view, the transition probabilities between two states $\rho_{1,2} \Rightarrow \Psi_{1,2}$ with reference to the equilibrium state, namely, $\rho_0 \Rightarrow \Psi_0$ (due to an underlying potential perturbation ϕ), are equal in both directions because they are proportional, respectively, to the two sides of the equation given by [96]:

$$|<\overline{\Psi}_0|\phi|\Psi_{12}>|^2 = |<\overline{\Psi}_{12}|\phi|\Psi_0>|^2$$

(7.40)

which is true inasmuch as ϕ is Hermitian. That is, in the case of neuronal dynamics, only with the possibilities of $i_2 = z(i_1)$ and $i_1 = z(i_2)$ the microscopic reversibility is assured; and then there would be a natural tendency for the microscopic parameter $\rho_{1,2}$ to move near to ρ_0.

7.7 Lattice Gas System Analogy of Neural Assembly

Consistent with the wave-functional characterization, the neuronal transmission can be considered as the interaction of a wave function with the (neuronal) assembly constituted by the repeated translation of the basic unit (neuron) cell. Such an assembly resembles or is analogous to a lattice structure in space comprising a set of locales (or cells) normally referred to as the *sites* and a set of interconnections between them termed as *bonds*. Relevant to this model, the neuronal assembly is regarded as *translationally invariant*. As a consequence, each cell (or site) is like any other in its characteristics, state, and environment. This refers to the equilibrium condition pertaining to the density of neuronal states. The implication of this translational invariance is the existence of extended states (or *delocalized states*). Further, the neuronal assembly can be regarded as a disordered system wherein the spatial characteristics are random so that the potentials (at each cell) are localized.

In the case of delocalized states inasmuch as all site energies are identical, the neuronal transmission refers to a simple case of the wave proliferating over the entire neural assembly; whereas, the localized situation would make some regions of the neuronal lattice more preferable (energy-wise) than the others so that the neuronal spatial spread is rather confined to these regions.

Considering the neuronal wave function localized on individual sites, the present problem refers to calculating the probability that the neural transmission occurs between two sites i and j with the transition resulting in the output being at one of the two dichotomous limits, namely, $+S_U$ or $-S_L$. This would lead to the calculation of the number of such transitions per unit time.

In the localized regime, the neuronal transmission from site i to site j (with energies E_i and E_j) corresponds to a transition probability w_{ij}. If the energy gaps between $+S_U$ and $-S_L$ states of ith and j^{th} units are ΔE_i and ΔE_j respectively, then regardless of the previous state set, $S_k = +S_U$ with probability $w_k = 1/[1 + \exp(-\Delta E_k)/k_B T]$, where k = i or j. This local decision rule in the neuronal system ensures that in thermal equilibrium, the relative probability of two global states is determined solely by their energy difference, dictated by the (pseudo) Boltzmann factor, namely, $\exp[-(E_j - E_i)/k_B T]$. This also refers to the extent of transmission link (weighting factor) between the sites i and j.

The corresponding intrinsic transition rate (from site i to site j) can be written as: $\gamma_{ij} = \exp[-2\alpha_i x_{ij} - (E_j - E_i)/k_B T]$, if $E_j > E_i$; or, $\gamma_{ij} = \exp[-2\alpha_i x_{ij}]$ if $E_j < E_i$ where α_i is specified by a simple ansatz for the wave function $\Psi_i(x_i)$ localized at x_i, taken as $\Psi_i(x_i) = \exp(-\alpha_i |x - x_i|)$ similar to the tunneling probability for the overlap of the two states; and $x_{ij} = |x_j - x_i|$. Further, the average transition rate from i to j is: $\Gamma_{ij} = <(M_i/M)(1 - M_j/M) \gamma_{ij}>$ where the M_i's are the neural cell population participating in the transition process out of the total population M. Assuming the process is stationary, Γ_{ij} can also be be specified in terms of the probability distribution function ρ_i and ρ_j as a *self-consistent approximation*. Hence, $\Gamma_{ij} = \rho_i(1 - \rho_j)\gamma_{ij}$. Under equilibrium conditions, there is a *detailed balance* between the transitions i to j and j to i as discussed earlier. Therefore, $(\Gamma_{ij})^o = (\Gamma_{ji})^o$, with the superscript o again referring to the equilibrium condition. Hence, $\rho_i^o (1 - \rho_j^o)(\gamma_{ij})^o = \rho_j^o (1 - \rho_i^o)(\gamma_{ji})^o$. However, $(\gamma_{ij})^o = (\gamma_{ji})^o \exp[-(E_j - E_i)/k_B T]$ which yields the solution that:

$$\rho_i^o = \{\exp[-(E_i - \phi_{pB})/k_B T] + 1\}^{-1} \tag{7.41}$$

and

$$\rho_i^o(1 - \rho_j^o) = \rho_i^o \rho_j^o \exp[-(E_j - \phi_{pB})/k_B T] \tag{7.42}$$

where ϕ_{pB} is the cellular (local) barrier potential energy (or the site *pseudo-Fermi* level.)

The Hamiltonian corresponding to the neuronal activity with the dichotomous limits ($+S_U$ and $-S_L$) corresponding to the possible interactions at i^{th} and j^{th} sites is given by Equation (7.28). Suppose the bias θ_i is set equal to zero. Then the Ising Hamiltonian (Equation 7.28) has a symmetry. That is, it remains unchanged if i and j are interchanged; or for each configuration in which a given S_i has the value $+S_U$, there is another dichotomous value, $-S_L$, such that $+S_U$ and $-S_L$ have the same statistical weight regardless of the pseudo-temperature. This implies that the neuronal transition ought to be zero in this finite system. Hence, within the framework of the Ising model, the only way to obtain a nonzero spontaneous transition (with the absence of external bias) is to consider an infinite system (which takes into account implicitly the thermodynamic limit). Such a limiting case corresponds to the classical continuum concept of wave mechanics attributed to neuronal transmission depicting the McCulloch-Pitts logical limits wherein the neuron purports to be an idealization of a real neuron and has features of being able to be excited by its inputs and of giving a step output (0 or 1) when a threshold is exceeded.

7.8 The Average Rate of Neuronal Transmission Flow

The weighting factor or the connectivity between the cells, namely, W_{ij} of Equation (7.28) is a random variable as detailed in Chapter 6. The probabilistic attribute(s) of W_{ij} can be quantified here in terms of the average transition rate of the neuronal transmission across the ith and j^{th} sites as follows.

The site energies E_i and E_j pertaining to the ith and j^{th} cells are assumed as close to the cellular barrier potential (or *pseudo-Fermi level*) ϕ_{pB}; and, the following relation(s) are also assumed: $|E_i|$, $|E_j|$, $|E_j - E_i|$ $\gg k_B T$. Hence, $\rho_i^o \approx 1$ for $E_i < 0$ and $\rho_i^o \approx \exp[-(E_i - \phi_{pB})/k_B T]$ for $E_i > 0$; and under equilibrium condition:

$$(\Gamma_{ij})^o = \exp\{-2\alpha_i \, x_{ij} - [|E_i| + |E_j| + (E_j - E_i)/k_B T]\} \qquad (7.43)$$

The corresponding neuronal transmission across i^{th} and j^{th} cells can be specified by a flow rate, ϑ_{ij}, equal to $\vartheta_{ij} = (\Gamma_{ij} - \Gamma_{ji})$. Now, suppose the perturbations at the equilibrium are $E_{ij} = (E_{ji})^o + \Delta E_{ij}$, $\rho_{ij} = (\rho_{ij})^o + \Delta\rho_{ij}$, and $\gamma_{ij} = (\gamma_{ij})^o + \Delta\gamma_{ij}$, and assuming that the detailed balance relation, namely, $(\Gamma_{ij})^o = (\Gamma_{ji})^o$ is satisfied, the following relation can be stipulated:

$$\vartheta_{ij} = \{[\Delta\rho_i/\rho_i^o(1 - \rho_i^o)] - [\Delta\rho_j/\rho_i^o(1 - \rho_j^o)] + [\Delta\gamma_{ij}/\gamma_{ij}]$$

$$- [\Delta\gamma_{ji}/\gamma_{ji}]\}(\Gamma_{ji})^o \qquad (7.44)$$

For a differential change in the local potential barrier energy $\Delta\phi_{pB}^i$ (at the pseudo-Fermi level), $\rho_i = 1/\{\exp[E_i - \Delta\phi_{pB}^i/k_B T] + 1\}$. Again, with $\Delta\phi_{pB} << k_B T$, ρ_i simplifies as equal to $\rho_i^o[1 + \Delta\phi_{pB}^i(1 - \rho_i^o)/k_B T]$ approximately with $\vartheta_{ij} = [(\Gamma_{ij})^o/k_B T] [\Delta\phi_{pB}^i - \Delta\phi_{pB}^i]$. This relation concurs implicitly with the observation of Thompson and Gibson [37,79] who observed that neuronal firing statistics "depends continuously on the difference between the membrane potential and the threshold". Further, it is evident that the neuronal transmission rate is decided by: (1) Rate of state-transition between the interconnected cells; and (2) the difference in the local barrier potentials at the contiguous cells concerned.

The parameter ϑ_{ij} is an implicit measure of the weighting factor W_{ij}. Inasmuch as the intracell disturbances can set the local barrier potentials at random values, it can be surmised that ϑ_{ij} (and hence W_{ij}) can be regarded as a nonstationary stochastic variate as discussed in Chapter 6.

Further, the extent of the cluster of cells participating in the neuronal transmission can be specified by the average range of interaction equal to $<x_{ij}>$ and the "size" of the wave packet emulating the neuronal transmission can be decided as follows: Suppose $(\Delta x_{ij})^{in}$ is the spatial spread of the incident wave at the site i. After the passage through the cell (local interaction site), the corresponding outgoing wave has the spread given by:

$$[(\Delta x_{ij})^{out}]^2 \approx [(\Delta x_{ij})^{in}]^2 + [\Delta v_{ij}]^2 [\tau]^2 \qquad (7.45)$$

where τ_{ij} is the ith to j^{th} site transit time.

7.9 Models of Peretto and Little *versus* Neuronal Wave

By way of analogy with statistical mechanics, Little [33] portrayed the existence of persistent states in a neural network under certain plausible assumptions. Existence of such states of persistent order has been shown directly analogous to the existence of long-range order in an Ising spin system, inasmuch as the relevant transition to the state of persistent order in the neurons mimics the transition to the ordered phase of the spin system.

In the relevant analysis, Little recognizes the persistent states of the neural system being the property of the whole assembly rather than a localized entity. That is, the existence of a correlation or coherence between the neurons throughout the entire interconnected assembly of a large number of cells (such as brain cells) is implicitly assumed. Further, Little has observed that considering the enormous number of possible states in a large neural network (such as the brain) — of the order of 2^M (where M is the number of neurons of the order 10^{10}) — the number of states which determine the long-term behavior is, however, very much smaller in number. The third justifiable assumption of Little refers to the transformation from the uncorrelated to the correlated state in a portion of, or in the whole, neuronal assembly. Such transformation can occur by the variation of the mean biochemical concentrations in these regions, and these transformations are analogous to the phase transition in spin systems.

On the basis of the above assumptions, Little has derived a $(2^M \times 2^M)$ matrix whose elements give the probability of a particular state $|S_1, S_2, ..., S_M>$ yielding after one cycle the new state $|S_1', S_2', ..., S_M'>$. (The primed states refer to the row, and the unprimed set, to the column of the element of the matrix.) This matrix has been shown analogous to the partition function for the Ising spin system.

It is well known that the correlation in the Ising model is a measure of the interaction(s) of the atomic spins. That is, the question of interaction refers to the correlation between a configuration in row q, say, and row r, for a given distance between q and r; and, when the correlation does exist, a long-range order is attributed to the lattice structure. For a spin system at a critical temperature (Curie temperature), the long-range order sets in and the system becomes ferromagnetic and exists at all temperatures below that.

Analogously considering the neuronal assembly, existence of a correlation between two states which are separated by a long period of time is directly analogous to the occurrence of long-range order in the corresponding spin system.

Referring to the lattice gas model of the neuronal assembly, the interaction refers to the incidence of neuronal wave(s) at the synaptic junction from the interconnected cells. The corresponding average range of interaction between say, the ith and j^{th} cell, is given by $<x_{ij}>$. The passage of a wave-packet in the interaction zone with minimum spread can be considered as the neuronal transmission with a state-transition having taken place. This situation corresponds to the spread of an incident wave packet, namely, $(\Delta x_{ij})^{in}$ equal to $[<x_{ij}>\bar{h}/p]^{1/2}$; and a zero spread can be regarded as

153

analogous to spontaneous transition in a spin-glass model. It represents equivalently the McCulloch-Pitts (logical) neuronal transition.

The presence of external stimuli (bias) from a kth (external) source at the synapse would alter the postsynaptic potential of the ith neuron. Little [33] observes that the external signals would cause a rapid series of nerve pulses at the synaptic junction; and the effect of such a barrage of signals would be to generate a constant average potential which can transform the effective threshold to a new value on a time-average basis. He has also demonstrated that this threshold shift could drive the network or parts of it across the phase boundary from the ordered to the disordered state or *vice versa*. That is, the external stimuli could play the role of initiating the onset of a persistent state representing the long-term memory.

In terms of wave mechanics considerations, the effect of external stimuli corresponds to altering the neuronal transmission rate; and, in the presence of external bias, ϑ_{ij} can be written as $\vartheta_{ij}' = [(\Gamma_{ij})^o/k_BT] [\Delta\phi_{pB}^i - \Delta\phi_{pB}^j]'$ where the primed quantity refers to the new threshold condition for the state transition.

The above situation which concurs with Little's heuristic approach is justifiable since the active state proliferation or neural transmission is decided largely by the interneuronal connection and the strength of the synaptic junctions quantified by the intrinsic state-transition rate γ_{ij}; and the external bias perturbs this value *via* an implicit change in the local threshold potential(s). The eigenstates which represent the neuronal information (or memory storage at the sites) warrant an extended state of the sites which is assured in the present analysis due to the translational invariancy of the neuronal assembly presumed earlier.

Little's model introduces a markovian dynamics to neuronal transmission. That is, the neurons are presumed to have no memory of states older than a specific time (normally taken as the interval between the recurring neuronal action potentials). Corresponding evolution dynamics has been addressed by Peretto [38], and it is shown that only a subclass of markovian processes which obeys the detailed balance principle can be described by Hamiltonians representing an extensive parameter for a fully interconnected system such as a neuronal assembly.

This conclusion in the framework of the present model is implicit due to the fact the intrinsic transition rate of a wave functional attribute prevails under equilibrium conditions with the existence of a detailed balance between the interconnection i to j or j to i sites.

7.10 Wave Functional Representation of Hopfield's Network

Consider a unit (say, m^{th} neuron) in a large neuronal assembly, which sets up a potential barrier ϕ_{pB} over a spatial extent a_{pB}. Assuming the excitatory situation due to the inputs at the synaptic node of the cell, it corresponds to the neuronal wave transmission across this m^{th} cell, with $C_{Tn} \approx 1$. The corresponding output or the emergent wave is given by the solution of the wave equation, namely, Equation (7.29) with appropriate boundary conditions. It is given by:

$$\Psi_{out}(m) = \Phi(m) \exp[j\{k(m)x(m) + e(m)\}] \tag{7.46}$$

where $k(m)$ is the propagation vector, $E(m)$ is the incident wave energy, $\Phi(m)$ is the mth mode amplitude function, $e(m) = [\pi E(m)a_{pB}/\lambda(m)\phi_{pB}]$ and $\lambda(m) = 2\pi/k(m)$.

Hence, the net output due to the combined effect of all the interconnected network of M neuron units at the mth synaptic node can be written as a superposition of the wave functions. That is:

$$\Psi_{Total}(m) = \sum_{\ell=1}^{M} \Phi(m-\ell)[\sigma(\ell)r(\ell)]\exp\{j[k(\ell)x(m-\ell) + e(\ell)]\} \tag{7.47}$$

where $\ell = 1, 2, ..., M$, and $\sigma(\ell)$ represents the incident wave at the synaptic node with a dichotomous value as dictated by its source/origin. That is, $\sigma(\ell) = 1$ refers to such a wave being present and $\sigma(\ell) = 0$ specifies its absence. Let the probability that $\sigma(\ell) = 1$ be w and the probability that $\sigma(\ell) = 0$ be $(1 - w)$. The parameter $r(\ell)$ in Equation (7.47) is a zero-mean, white gaussian sequence which depicts the randomness of the stochastical inputs at the synaptic summation. Further, Equation (7.47) represents a simple convolution process which decides the neuronal input-output activity under noise-free conditions.

Suppose intraneuronal disturbances are present. Then a noise term should be added to Equation (7.47). In terms of wave function notations, this noise term $\eta(m)$ can be written as: $\eta(m) = \Phi_{\eta}(m)\exp[j(\xi_{\eta}(m))]$, where the amplitude Φ_{η} and the phase term ξ_{η} are random variates, (usually taken as zero-mean gaussian). Hence, the noise perturbed neural output can be explicitly specified by:

$$\Psi_{Total}^{\eta}(m) = \sum_{\ell=1}^{M} \Phi(m-\ell)[(\sigma(\ell)r(\ell))\exp[j\{\xi(m)\}]$$

$$+ \Phi_{\eta}(m-\ell)]\exp[j\{\xi_{\eta}(\ell)\}] \tag{7.48}$$

where $\xi(m) = [k(\ell)x(m-\ell) + e(\ell)]$, and $\xi_{\eta}(\ell) = [\pi E_{\eta}(\ell)a_{pB}/\lambda(\ell)\phi_{pB}]$, with $E_{\eta}(\ell)$ depicting the eigen-energy associated with the noise or disturbance.

The nonlinear operation in the neuron culminating in crossing the threshold of the potential barrier corresponds to a detection process decided by the input (random) sequence $r(m)$ so that the summed input exceeds the barrier energy across the neuron. Such a detection process refers to minimizing the mean square functional relationship given by:

$$\varepsilon = 1/2 \sum_{m=1}^{M} [\Psi_{Total}^{\eta}(m) - \Psi_{Total}(m)]^2 \qquad (7.49)$$

Written explicitly and rearranging the terms, the above relation (Equation 7.49) simplifies to:

$$\varepsilon = -1/2 \sum_{m=1}^{M} \sum_{n=1}^{M} W(m, n)\sigma(m)\sigma(n) - \sum_{m=1}^{M} \theta(m)\sigma(m) \qquad (7.50)$$

with $W(m, m) = 0$ and $m, n \in 1, 2, ..., M$; further:

$$W(m, n) = -\sum_{\ell=1}^{M} \Phi(\ell - m)\Phi(\ell - n)r(m)r(n) \qquad (7.51)$$

and

$$\theta(m) = \sum_{\ell=1}^{M} [\Phi(\ell - m)r(m)\Psi_{Total}^{\eta}(m)] - (1/2)[|\Phi(\ell - m)|^2|r(m)|^2] \qquad (7.52)$$

The ε of Equation (7.50) depicts a neural network with the weights of interconnection being W and an external (bias) input of θ. Thus, the energy function of the Hopfield network can be constructed synonymously with wave functional parameters.

7.11 Concluding Remarks

The application of the concept of wave mechanics and the use of quantum theory mathematics in neurobiology were advocated implicitly by Gabor as early as in 1946. As stated by Licklider [75], "the analogy [to] the position-momentum and energy-time problems that led Heisenberg in 1927 to state his uncertainty principle ... has led Gabor to suggest that we may find the solution [to the problems of sensory processing] in quantum mechanics." Supplemented by the fact that statistical mechanics too can be applied to study the neuronal activity, the foregoing analyses considered can be summarized as follows:

The neuronal activity can be represented by the concepts of wave mechanics. Essentially, considering the fact that the interconnected neurons assume randomly one of the dichotomous potentials (0 or ϕ_{pB}), the input sequence at any given neuron would set a progression of state transitions in the interconnected cells. Such a spatial progress or the "collective movement" of state-transition flux across the neuronal assembly can be regarded as the neuronal transmission represented as a wave motion.

Hence, the dynamic state of neurons can be described by a set of extensive quantities *vis-a-vis* the wave functional attributions to the neuronal transmission with relevant alternative Hamiltonian perspectives presented.

Accordingly, the neuronal transmission through a large interconnected set of cells which assume randomly a dichotomous short-term state of biochemical activity can be depicted by a wave equation.

In representing the neuronal transmission as a "collective movement" of neuronal states, the weighting factor across the neural interconnections refers implicitly to a long-term memory activity. This corresponds to a weight space Ω with a "connectivity" parameter (similar to the refractive index of optical transmission through a medium) decided by the input and local energy functions.

The wave mechanical perspectives indicate that the collective movement of state transitions in the neuronal assembly is a zero-mean stochastical process in which the random potentials at the cellular sites force the wave function depicting the neuronal transmission into a *nonself averaging* behavior.

Considering the wave functional aspects of neuronal transmission, the corresponding eigen-energies (whose components are expressed in terms of conventional wave parameters such as the propagation constant) can be specified.

The wave mechanical considerations explicitly stipulate the principle of detailed balance as the requisite for microscopic reversibility in the neuronal activity. Specified in terms of the strength of synapses, it refers to $W_{ij} = W_{ji}$. This symmetry condition restricts the one-to-one analogy of applying the spin-glass model only to a limited subclass of collective processes.

The neuronal assembly can also be regarded as analogous to a lattice gas system. Such a representation enables the elucidation of the probability of state transitions at the neuronal cells. That is, by considering the neuronal assembly as a disordered system with the wave function being localized, there is a probability that the neural transmission occurs between two sites i and j with the transition of the state $+S_U$ to $-S_L$ (or *vice versa*) leading to an output; and hence the number of such transitions per unit time can be specified.

In terms of the wave mechanics concept, the McCulloch-Pitts regime refers to the limit of the wavelength being so very short (with $\phi_{pB} \ll E_i$) that any physically realizable potential would change only by a very negligible amount assuring a complete transmission. Within the framework of the Ising spin model, such a nonzero spontaneous transition would, however, warrant an infinite system (in the thermodynamic limit and assuming the external bias being zero) as observed by Peretto.

In the existing studies based on statistical mechanics modeling of neuronal activity, Peretto identifies an extensive parameter (expressed as a Hamiltonian) for a fully interconnected Hopfield network, and relevant state-transitional probability has hence been deduced. The basis for Peretto's modeling stems from the existence of a long-range persistent order/state in biological neurons (analogous to or mimicking the Ising spin system), as observed by Little.

Peretto, by considering the output of action potentials occurring at regular intervals due to the synchronized excitatory (or inhibitory) synaptic action, has elucidated the markovian aspects of neuronal activity. This is substantiated by the evolution equation of the system described in Chapter 6. Peretto deduced a digital master equation to characterize the markovian structure of neuronal transmission. He has indicated the existence of a Hamiltonian at least for a narrow subclass of markovian processes which obey the detailed balance principle governing the state-transition rate. Similar observations are plausible by considering the Fermi golden rule as applied to the state-transition probability and its dynamics governed by Boltzmann's equation. It is inferred thereof that regardless of the microscopic origin of neuron interactions, the principle of detailed balance must be satisfied in the neuronal dynamic process.

Consideration of Boltzmann's equation indicates that the probability of neurons being in the equilibrium state will decay exponentially. From the neural network point of view, it implicitly refers to the well-known integration process (low-pass action) associated with neuronal input/output relation.

The average rate of neuronal transmission flow depends on the rate of state-transition between the interconnected cells and the difference in the local barrier potentials at the contiguous cells concerned. This is in concurrence with a similar observation by Thompson and Gibson. The average rate of transmission has also been shown as an implicit measure of the weighting factor. Further, by considering the spatial "spread" or the "size" of the wave-packet emulating the neuronal transmission, the spread when minimum represents the passage of a wave-packet across the cell. In other words, it represents the neuronal transmission with a state-transition having taken place.

In terms of wave mechanics considerations, the effect of external stimuli refers to altering the neuronal transmission rate. This is in concurrence with Little's heuristic justifications that the active state proliferation is decided largely by the interneuronal weighting function. Any external bias would perturb this value *via* an implicit change in the local potentials.

The eigenstates of the "neuronal wave" represent the neuronal information (or memory storage at the sites). The existence of eigenstates presumably warrants an extended state of the cellular sites which is guaranteed by the translational invariancy of the neuronal assembly.

Considering the presence of intraneural disturbances in a Hopfield network the corresponding system can be modeled in terms of wave functional parameters as could be evinced from Equation (7.50).

The spread in the neuronal wave function across a cell is an implicit indicator of whether state transition has occurred or not. The minimum spread assures the transmission of the neuronal wave confirming that transition has occurred; and its zero value (ideally) refers to the McCulloch-Pitts logical transition or the spontaneous transition.

This spread in the wave function can be equated to the"smearing" condition proposed by Shaw and Vasudevan. That is, from the thermodynamics point of view the neuronal transition (when modeled as analogous to the Ising principle of interacting spins), corresponds to the McCulloch-Pitts model with the $k_B T$ term tending to zero. This would require, however, a total absence of fluctuations in the postsynaptic potentials; but, in the real neurons, there is an inevitable fluctuation in the summed up postsynaptic potentials, which leads to a finite spread in the wave-functional transmission.

The amplitude of the transmitted wave function, namely, $C_{Tn} a(\Omega_i/\Omega)$ is an implicit function of the domain fraction (Ω_i/Ω). The subset(s) of Ω_i across which a preferential wave transmission occurs decide the so-called persistent order (based on a learning mechanism) attributed to neurons by Little.

The quasiparticle and wave-like propagation of neural transmission describe a kinetic view that stems from a phase space kinetic equation derived from the wave equation. The resulting field theory picture of neurons is a wave train corresponding to a system of quasiparticles whose diffusive kinetics permits the elucidation of the amplitude and phase properties of the propagating wave train in a continuum and thereby provide a particle portrait of neuronal dynamics. Griffith once observed the "concept of nervous energy is not theoretically well-based, at least not yet ... However, although there are obstacles it may be argued they are not necessarily insurmountable."

CHAPTER 8

Informatic Aspects of Neurocybernetics

8.1 Introduction

It was discussed in the previous chapters that a neural network is a conglomeration of several subsets which are in part its constituents and in part interact with it through massive interconnections. Also was indicated that the subspaces of a neural domain have feedbacks portraying explicit influence of one subsystem over its neighbors. A neural network represents a self-controlling set of automata, a memory machine, and a homeostat. It is a domain of information-conservation — a process in which a chain of elements (cells) takes part in the collection, conversion, storage, and retrieval of information.

The neural system is also amenable for isolation into individual subsets such as cells, interconnections, synapses, etc.; and each subsystem may in turn be broken down into microscopic structures (parts) participating in the neural activity.

Thus, a neural network represents a complex automation — a *cybernetic system* [9]. The association of its subsystems *via* feed-forward and/or feedback constitutes an optimum control for self-organization. As a science, neurocybernetics deals with the operation of an automaton and its characteristic as an integral self-controlled system seeking an optimum performance. In this endeavor, for the efficient use of methods enabling cybernetically controlled (self-adapting) activities, the neural system deliberates the minimization of uncertainties arising from the inherent noise or (spatiotemporal) random characteristics of its activities.

Such activities refer to the state-space of the units (cells) comprising the neural complex. Compilation of data on the state of the system, transmission of the data (among the interacting units), and storage or retrieval of data where and when needed constitute the information-processing tasks in the neural system. Pertinent to neurocybernetics, information can be in a wider sense defined as a *measure of the disorganization* removed (a *counter default measure*) from the cellular complex by receiving knowledge which is *pragmatically* utilized as actions of self-organizational procedures. In this perspective, an informative entity is *usefully* implementable in the neurocybernetic processes. That is, not only it does have a representative value available to depict a disorganization, but also it lends itself to effective actions by existing means for self-organization.

To realize optimum self-organizing characteristics, the neural system warrants a minimum amount of information. The associated disturbances in the neural complex decide the extent of uncertainties in the neuronal states, thereby augmenting the system entropy which influences the amount of information need to be processed. The minimum information required is, therefore, implicitly assessed *via* entropy considerations. The question of minimum information arises because an information processor, in general, could be nonideal and lose information (*null-information*) as well as gain *false* and/or *redundant* information.

In real (biological) neural system, *mossy fibers* which originate from several brain sites carry information of diverse nature concerning the *declarative* aspects of internal states (of the interconnected neurons) and the *descriptive* message about the physioanatomical environment. The items of information so transmitted or delivered as an input to the neural communication network are largely sensory data. In addition, there are also information-bearing commands of the central motor system which coexist in the interneural communication links.

The information discharge rates of mossy fibers are modulated over a wide range which permits them to transmit detailed parametric information. Therefore, information-theoretic analyses could be done in a domain which represents broadly a *parametric spread-space.*

The self-sustaining activity of the neural network arises from the dynamics of neural mechanisms and is due to feedbacks of recurrent circuits in the interconnected complex. Further, such self-regulating/sustaining activities refer to the generation of a mode (activity) pattern performed by cerebellar modules which are matched with patterned outputs, and realization of such matching corresponds to achieving an *objective goal (function)*. The self-regulation process is, therefore, an adjunct accomplishment of the associative memory problem, namely, the network's response in delivering one of the stored patterns most closely resembling that which is presented to it. The closeness of resemblance approaching infinity corresponds to the realization of the objective function.

8.2 Information-Theoretics of Neural Networks

Application of informational analysis to the neural complex has so far been limited to the information capacity heuristics of structures like Hopfield's net model of associative memory — for the purpose of quantifying such network performance as a memory. The information capacity of a standard memory is determined explicitly by the number of memory bits in simple models, but where complex dynamics of activation patterns are encountered, probabilistic estimates of the information capacity are deduced on the basis of certain simplifying assumptions. For example, estimates of the capacity of Hopfield's network *via* statistical methods have been formulated and extended with more rigorous statistical techniques [97-100]. Further, Lee et al. [101] improvised an approach using *Brown's Martingale Central Limit Theorem* and *Gupta's Transformation* to analyze the complex dynamics of Hopfield's memory model rigorously through information-theoretic considerations. As an extended effort, Gardner and Derrida [102,103]

161

considered the relevant aspects of storage capacity of neural networks in search of optimal values. Pertinent to neurons, the *memory* and *thought-process* are decided by the associative properties of the collective neuronal aggregates and the underlying entropy considerations. In this perspective, Caianello [64] outlined a theory of thought-process and *thinking machines* in terms of *neuronic equations* depicting the instantaneous behavior of the system and *mnemonics equations* representing the permanent or quasi-permanent changes in the neuronal activities. Such permanent changes in neural functioning (caused by experience) have been modeled as a degree of plasticity which decides the *fixation of memory* or *memory storage* involving time-dependent processes.

Shannon's concept of information or entropy [104,105] in stochastic systems has also been extended to neuronal associative memory which has led to the information-theoretic modeling of neural activity manifesting as a neuronal potential train of spikes. Accordingly, the temporal and spatial statistics of this neuronal aggregate of signal elements have featured dominantly in the information-theoretic approaches formalized in the 1970's [106]. For example, Pfaffelhuber in 1972 [107] used the concept of entropy to describe the *learning process* as one in which the entropy is depicted as a decaying function of time.

The concept of entropy has also formed the basis of elucidating the *information capacity* of neural networks. For example, considering the associative memory as a plausible model for biological memory with a collective dynamic activity, the information capacity of Hopfield's network has been quantitatively elucidated.

Classical characterization of information in a neural network *via* stochastic system considerations (as conceived by Shannon and Weaver) simply defines a specification protocol of message transmission pertinent to the occurrence of neuronal events (firings) observed (processed) from a milieu of possible (such events) across the interconnected network. Thus in early 1970's, the information processing paradigms as referred to the neural system were derived from the stochastic considerations of (or random occurrence of) neural spike trains; relevant studies were also devoted to consider the probabilistic attributes of the neural complex; and, considering the randomly distributed cellular elements, the storage capacity of information in those elements (or the associated memory) was estimated. That is, the information-theoretic approach was advocated to ascertain the multiple input-single output transfer relation of information in the neuronal nets. The information capacity of a neural network was then deduced on the basis of the dynamics of activation patterns. Implicitly, such a capacity is the characteristic and ability of the neural complex viewed as a collective system. It was represented as a memory (that stores information) and was quantified *via Hartley-Shannon's law* as the logarithm of number of strings of address lines (consisting of memory locations or units) distinguished. Memory locations here refer to the number of distinguishable threshold functions (state-transitional process) simulated by the neurons.

In the elucidation of optimal (memory-based) storage capacity of neural networks, the following are the basic queries posed:

- What is the maximum number of (pattern) examples that the neural complex can store?
- For a given set of patterns (less than the maximum value), what are the different functions that could relate the network inputs to the output?
- How do the statistical properties of the patterns affect the estimation of the network information capacity?

Further, the notions of information capacity of the neural complex (associated memory) are based on:

- Binary vector (dichotomous) definition of a neuronal state.
- Matrix representation of the synaptic interconnections.
- Identifying the stable state of a neuron.
- Defining the information capacity of the network as a quantity for which the probability of the stored state-vector patterns (as per Hebbian learning) being stable, is maximum.

Information flow across the real neural complex, in general, often faces an axonal bottleneck. That is, the input arriving to a neuronal cell is not as information poor as the output. On the input-end neurons are more ajar than on the output-end. Often they receive information at a rate three orders of magnitude higher than they give it off. Also, the neurons always tie together to form a group and, hence, permit an anisotropic proliferation of information flow across the neural complex. These groups were termed as *compacta* by Lengendy [108]. It is the property of compacta (due to massive interconnections) that, whenever a compactum fires, the *knowledge* imparted by the firing stimulus is acquired by every neuron of the compactum. Thus, neuronal knowledge proliferates from compactum-to-compactum with a *self-augmentation* of information associated in the process. That is, each successive structure is able to refine the information available to it to a maximum extent that the information can possibly be refined. Due to this self-augmentation of information, McCulloch and Pitts called the neural nets as *networks with circles*. Further, routing of information across the interconnected compacta is a *goal-pursuit problem*. That is, the self-organizational structure of the neural complex permits a goal-directed neural activity and the goal-seeking is again monitored adaptively by the associated feedback (self)-control protocols.

163

Pertinent to conventional memory, information storage is rather an explicit quantity. For example, in a Random Access Memory (RAM) with M address lines and 1 data line (2^M memory locations, each storing 1 bit of information), the storage capacity is 2^M bits. That is, the RAM as an entity distinguishes 2^{2^M} cases (in respect of setting 2^M bits independently as 0 or 1) and thereby stores a string of 2^M bits. This definition enables the entire contents of the RAM as one entity that encodes a message — the logarithm of such different messages measures the information content of the message.

Along similar lines, how can the capacity of a neural network be defined? The information content in a neural net refers to the state-transitions, the associated weights, and thresholds. How many different sets of weights and thresholds can then be distinguished by observing the state-transitions of the network? Abu Mostafa and St. Jacques [98] enumerated such threshold functions involved and showed that there are $2^{\alpha N^3}$ distinguishable networks of N neurons, where α is asymptotically a constant. In logarithmic measure, the capacity of feedback network is, therefore, proportional to N^3 bits. However, the above definition of capacity of a neural net is rather heuristic and not apparent. In order to elucidate this capacity explicitly, one has to encode information in the state-transitions across the interconnected network; and decoding of the message warrants observations concerning "which states go to which state". Such a format of depicting information storage is not, however, practical or trivial.

Alternatively, the information storage in feedback networks can be specified *vis-a-vis* the *stable states*. For this purpose, an energy function can be defined to indicate the state-transitions leading to the *stable state*.[*] The stable states are vectors of bits (that correspond to words in a regular memory), and convergence to such stable states is the inherency of a feedback network functioning as an associative memory. Now, how many such stable states can be stored in a feedback network? The number of stable states can be deduced as bN, each consisting of N bits (depicting the individual states of N neurons), and b is an asymptotic constant. Hence, stable state capacity of a feedback network of N neurons is proportional to N^2 bits. The reduction from N^3 to N^2 arises from the loss of information due to the restricted (selective) observations depicting only those transitions leading to stable states. The selection of stable states used in the computation of the memory storage is a rule-based algorithmic strategy that considers a set of vectors and produces a network in which these vectors are stable states. For example, the Hebbian rule chooses the matrix of weights to be the sum of outer products of the vectors to be stored. The stable-state storage capacity of the neural

---

[*] *Stable state* : It refers to a state where the neuronal state remains unchanged when the update threshold rule is applied.

network is dependent on the type of algorithmic rule chosen. Relevant to Hebbian rule, for example, only (gN/log N) randomly chosen stable states can be stored (where g is an asymptotic constant). The corresponding capacity is then proportional to $N^2/ \log N$ bits.

For a feed-forward network, there is no definition of memory capacity that corresponds to stable-state capacity inasmuch as such networks do not have stable states, but rather input-output relations only.

As discussed in Chapter 4, the neural complex has an architecture with layers of cellular units which are fully interconnected and each unit can be regarded as an information processing element. In reference to this layered architecture, Ackley et al. [57] used the concepts of reverse cross-entropy (RCE) and defined a *distance function* to depict the deviation of neural statistics in the presence of environmental inputs from that in the absence of such inputs in the entropy plane. Minimization of this distance parameter refers to the process of attaining the objective function. Both the reverse cross-entropy approach due to Ackley et al. and an alternative method based on cross-entropy concepts as proposed by Liou and Lin [58], describe implicitly an entropy-based, information-theoretic approach to analyze a layered framework of neural units.

The information aspects of a neural complex viewed in terms of memory capacity also penetrate the coordinated organization of firing patterns in recurrently connected cellular units. That is, the *memory traces* refer to systematic sequential firing of particular sets of neurons and the *variety* in memory space depicts the vagaries in traces pertinent to different sequences of different sets of neurons. In the universe of the neural complex, any cellular unit may, in general, participate in many such traces defined by the temporal relations among the participating neurons (as observed by Little [33]). The traces are embedded in a given cellular set *via* selective modulations of the synaptic weighting among the participant neurons in accordance with the fashion of arrangement (of the neurons) in sets in the firing patterns. Not only the temporal traces but also the spatial proliferation of state-transitions are the embodiments of memory considerations in the neural system. As observed by MacGregor [109], "from the very onset, anatomical connections and associated temporal firing patterns are two distinct but intrinsically coupled manifestations of a unitary phenomenon."

Do the memory traces overlap? MacGregor [109] answers this question by considering the *cross-talk* mediated by inappropriately activated synapsis through synaptic adjustments. That is, two distinct traces may call for the same single synapse between two cells that participate in both traces, but each asks for a distinct value to be assigned to the synapse; or it is likely in multiple embedded nets that a given neural cell may project to some subset of cells because of its position in one trace and to another subset of cells as a result of its position in a second trace. Implicitly, this is the same as Gardner and Derrida's approach [102,103] of learning an explicit target function or extraction of a rule formalized in terms of replica symmetry consideration discussed in Chapter 4.

The multiple traces embedded in a random fashion can be regarded as being completely independent of each other, and the cross-talk can be regarded as a disruptive aspect of memory borne by a recurrently connected network. The stochastical theory of cross-talk has been comprehensively addressed by MacGregor who associates the concepts of *beds* and *realizations* in describing the memory traces; that is, the firing of specific sequences of sets of neurons which represent the items of information consists of physiological variations (or realizations) of primarily anatomical sites which are dubbed as beds. Thus, the population of excitatory and inhibitory neurons taken as an ordered sequence of sets constitutes the bed of a trace, and a realization is an observable physiological manifestation of an underlying bed. It consists of an ordered sequence of subsets of active neurons some of which are members of the corresponding sets of the bed and which fire over a given time interval in the temporal correspondence that exists among the cells of the bed.

Using the above notion in describing the neural complex, MacGregor and Gerstein [110] elucidated mathematical expressions to show how the memory capacity of recurrently connected nets depends on their characteristic anatomical and physiological parameters; and considerations of overlaps of traces, disruptive cross-talks, and random background activities were judiciously algorithmized in determining the storage capacities of information (*via* traces) pertinent to large interconnected networks.

The aforesaid existing body of information-theoretics as applied to the neural complex, however, does not *per se* determine the *pragmatic* and *semantic* utility of neural information with regard to self-organizing (cybernetic) control-goals. This is due to the fact that the techniques pursued by the *statistical theory of information* do not match the analysis of control problems. Classical statistical theory of information describes only the processes involved in the transmission and storage of information, but it is not comprehensible to treat information *vis-a-vis* control strategies or the extent of contraction of the parameter subspace as a result of previous training.

8.3 Information Base of Neurocybernetics

The neural complex handles information at three stages: The *input stage* (Figure 8.1), the *processor stage* , and the *controlling stage* (Figure 8.1). The *input information* includes all the *objective knowledge* on the structure, arrangement, and properties (such as the synaptic anatomy, physiology, and biochemical characteristics) of the participating neurons in the control endeavor. It also covers the details on the neural environment of the space (or domain) accommodating these neurons.

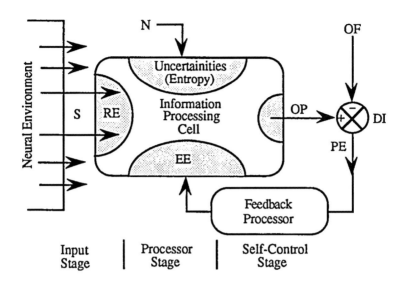

Figure 8.1 Information traffic in the neurocybernetic system
S: Sensory input information; RE: Synaptic (recognition) information of the
inputs; N: Intra- or extracellular disturbances (noise); OP: Axonal output
information; DI: Decision-making information; OF: Learning-based objective
function for self-control endeavors; PE: Predicted error information;
EE: Error-correcting information for self-organization

The *processing information* characterizes the relevant properties and
structure of the neural activity or state-transitional events across the
interconnected neurons (these include the excitatory and inhibitory aspects,
delays, threshold levels, inherent disturbances, and a variety of processing
neuronal infrastructures). In essence, processing refers to all relevancies of the
control center which strives to attain the objective function being fullfilled. It
is a *rule-based set of algorithms* [*] which extracts useful information (from the
inputs) and utilizes it to achieve the goal.

Once the information is processed, it represents a *controlling
information* which is looped (forward or backward) into the system to refine
the achievement of the goal or reduce the *organizational deficiency* that may
prevail and offset the efforts in realizing the *objective function*. That is, the
controlling information is the knowledge exerted to self-control (or regulate)
the automata. The controlling information process is therefore a stand-alone

---

[*] An *algorithm*, in general, refers to a sequence of steps or codes or rules
leading to specific results.

strategy by itself, distinctly operating on its own as an adjunct to the main information processor.

The controlling information includes morsels of knowledge associated with sensing and recognition of the processed information (from the second stage), algorithmic manipulations on the sensed data for decision making, and strategies of predicting the errors. It supplies information to neural actuator(s) to execute the feedback or feed-forward controls through the organizational loop. The controlling information-processing works towards the realization of objectives or the goals with the minimization of errors.

Thus, the concept of information-processing in a neural complex viewed from a cybernetic angle assumes the informational structure is pertinent not only to perceiving knowledge from the source and analyzing it for its usefulness, but also to processing it (further) for applications towards achieving a self-organizing automaton. Specific to the application of information theory to neurocybernetics, the following general considerations are therefore implicit:

• In view of the automatic and adaptive control strategies involved in neurocybernetics, an informatic-based *transfer function* should be defined which refers to information-processing algorithms concerning the attainment of a target or objective function; and also in terms of informational *efficiency functions* they should assess how a given processing algorithm is realized by the control strategies involved (subject to constraints on speed of response, storage capacity, range of inputs, number of interconnections, number of learning iterations, etc.).

• There should be elucidation of methods by which the informational characteristics of the neural network can be derived by those strategies such as frequency domain analysis normally employed in the theory of adaptive controls.

• There should also be evaluation of the effectual or ineffectual aspects of classical stochastical theory of information in describing (either quantitatively or qualitatively) the control activities of a neural network. In general, statistical information describes the quantitative aspect of the message and does not portray the qualitative consideration of the utility of the message in contributing the attainment of the objective function. Further, the operation of a neural system is dynamic; and the statistical measure of information is unfortunately inadequate to include the nonstatistical part of the dynamic system such as its topological, combinatorial, and algorithmic features.

• The cybernetic perspective of a neural network represents the *degree of organization* of neural activity. Therefore, the corresponding informatic description of the neural system should not only address the memory considerations, but also enclave the control aspects of modeling and programming of the collective response of neurons.

•   From a neurocybernetic viewpoint, the information theory should address the *semiotic aspects* of information, covering the *syntatics* which relate the formal properties of neurons and their combinations (variety) to the amount of information they carry; and *semantics* and *pragmatics*, which define the information content and the information utility of the neural signal elements (constituted by the binary state vectors of the cellular potential transitions), respectively.

•   By enunciating a relation between the degree of self-organization *versus* the informatics of *orderliness*, a new approach as applicable to neural cybernetics can be conceived. Wiener indicated such a trend as to incorporate and extend information theory to neural cybernetics *via* semiotic considerations.

•   The concept of neurocybernetic informational theory rests upon the *threshold of distinguishability* of neural state variables and the amount of their *variety* pertinent to the self-control process.

•   Analysis and synthesis of man-machine systems as done in the development of computer architecture and artificial neural networks mimicking the biological neurons refer to modeling and programming only at the information structural level. Such modeling and programming *via* information theory in the neurocybernetic domain should, however, broadly refer to:

a.   Informational description of the global neural complex.

b.   Similarity or dissimilarity criteria with regard to the objective structures, information structures, and information flows specified in terms of entropy parameters of the neural complex.

c.   Establishing a similarity between informational functions of processing by self-organizing (control) centers of the interconnected cells.

The bases of inculcating an informatic approach to neural cybernetics are the *complexity* of the system (in spatiotemporal domains), *orderliness* (alternatively randomness) associated with the system due to the presence of inevitable intra- and extracellular disturbances, *degree of organization* (of self-control) effected by feedback control strategies, and the *entropy* of the system [111].

The *complexity* of the neural system is decided basically by the number of units (cells), the variety of the functional events (state variables such as the energy levels), and the complexity of occurrence of such events in the time domain. An algorithmic description of the neural complexity in the informational plane should therefore include all the aforesaid considerations summarized by MacGregor [109] as beds and realizations.

Neural complexity is a generalized estimate functionally related to the *number variety* (composition), structure, and properties of the cellular units considered in space or time or both. The overall complexity could be nonadditive of the influences arising from the number of cellular units participating in the neural activity and their *variety*. For example, the excitatory neurons, the inhibiting neurons, the neurons with different threshold levels, the neurons with varying extents of synaptic inputs, etc. constitute the "variety" features indicated above. In other words, variety is an implicit attribution of diversed nature of beds and realizations constituting the items of neural informatic domain.

8.4 Informatics of Neurocybernetic Processes

The essential parameters deciding the informatic aspects of self-control functions in a neural network viewed in cybernetic perspective are:

- Complexity of neural architecture.
- Spatiotemporal randomness (disorder of the system) of neural state transitions.
- Self-organization efficiency of the interconnected cells.
- System entropy associated with the spatiotemporal neural events.

Corresponding informational analysis pertinent to the self-controlling or organizing characteristics of the neural system can be specified in terms of the following entities:

- Information processing algorithms at three stages of neural information traffic depicted in Figure 8.1
- Information utility or the value of neural information.
- Information efficiency resulting from loss of information due to intra- and/or extracellular disturbances.
- Steady and transient (dynamic) states of neural information flow.

A major function of a complex neurocybernetic system is a goal-related (dictated by an objective function), self-organizing (or self-regulating) effort viewed within a set of bounds. The associated randomness or disorderliness causes the system parameters (specified by a vector set) to veer from the system objective. The corresponding deviatory response in a neural network can be quantified by an ensemble of *diversion factors* pertinent to the neural environment which can be subdivided as follows:

- External (extraneural) diversion subset (due to disturbances or randomness introduced by external influences such as extracellular disturbances).
- Internal (intraneural) diversion subset (due to causes arising from internal randomness or cellular noise).

8.5 Disorganization in the Neural System

The task of self-control through adaptive feedback in the neural complex is to achieve a self-organization overcoming the influence of randomness which otherwise would promote a deviatory response from the objective or target response. The extent of disorganization so promoted can be generalized in terms of a *self-organization deficiency parameter*. That is, any disorganization perceived can be correlated to the randomness and the system complexity. Accordingly, the *self-organization deficiency* of a neural complex is defined as:

$$O_D = \left[\bigcup_{\alpha}^{\ell} \tau_\alpha \right] \left[\bigcup_{\beta}^{m} r_\beta \right] \left[\bigcup_{\gamma}^{N} W_\gamma \, d(Y_\gamma) \right] \tag{8.1}$$

where U represents the union operation in the spatiotemporal domains over ℓ instants of time, m situations (or realizations), and N cellular units (with τ, r, and W being the corresponding weights, respectively); and $d(Y_\gamma)$ denotes the randomness or *disorderliness function* associated with the γth neural unit as explained below.

Disorderliness is a measure of deviation of a selected variable, say y_j, with reference to a specified standard of order, y_T. Geometrically, if y_T refers to the target vector pinpointing the center of a region of orderliness, around this center a *quasi-ordered* region can be considered wherein the orderliness is maintained within a bound constrained by a (statistical) standard deviatory limit (Figure 8.2). The disorderliness can be assessed in terms of $D(y_j)$, the *distance* from the center of orderliness to the boundary of the quasi-ordered region. Therefore, the disorderliness can be written as [111]:

$$Y_j = |y_j - y_T| - D(y_j) \tag{8.2}$$

where $|y_j - y_T| = Q_j$ is the magnitude of the error vector. Y_j can be rendered dimensionless by normalizing it in terms of a reference (disorderliness) value. The disorganization function $d(Y_j)$ refers explicitly to the effect of disorderliness perceived at j^{th} unit of the neural system.

171

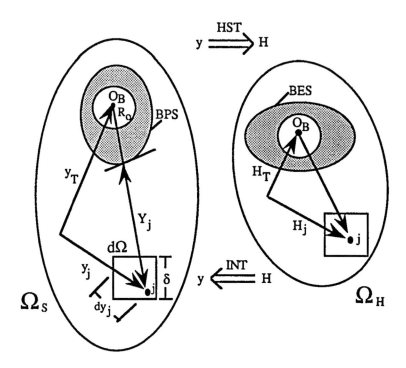

Figure 8.2 Parameter spread-space and entropy-space of the neural complex HST: Hartley-Shannon transform; INT: Inverse transform; O_B: Locale of the objective function; BPS: Boundary of the quasi-order of parameter spread space; BES: Boundary of the quasi-order region of entropy space

The orderliness (or the disorderliness) of the neural system is also influenced by the *system complexity* C_s which as observed from the system exterior refers to the variety of the system subsets and/or microsubsets. Pertinent to the neural complex, the mixture of cells with excitatory and inhibitory states and individual synaptic (physiochemical and/or anatomical) properties considered in the spatiotemporal domains (as beds and realizations) constitute typically the universe of *system complexity*. Mathematically, it can be represented by:

$$C_s = \phi(N, \upsilon) \tag{8.3}$$

where N is the number of cellular units and υ is their (associated) *variety* discussed earlier. The function ϕ measures implicitly the uncertainty or entropy due to the complexity and therefore is logarithmic as specified by *Hartley's law* .

The uncertainty indicator of the complexity should therefore be limited to the disorderliness function by a criterion that, in the event of the system tending to be well-ordered (minimization of disorderliness) with the condition $Y_j \to 0$, it should be accommodated by an appropriate functional relation between Y_j and C_s.

Further, by considering the geometrical space representing the disorderliness, the variate y defines a measure of uncertainty region (deterministic or probabilistic) and it also refers implicitly to the *a priori* probability of a sample within the region.

The collective influence of disorderliness and the complexity, therefore, determine the extent of entropy (or uncertainty) associated with the self-organizational efforts in the neural system in achieving a specific control goal. That is, the net complexity and disorderliness function bear the information on the degree of disorganization or the self-organization deficiency defined earlier.

If self-organizational deficiency is attributed to every synaptic coupling, across N cellular units, the overall self-organization deficiency can be stipulated by:

$$O_D = \sum_{j=1}^{N} p_j d_0(\Theta_{Yj}) \qquad (8.4)$$

where p_j refers to the probability of encountering the jth cell (in the neural spatial complex) wherein the disorganization is observed, d_0 is a function to be specified, and Θ_Y is a *disorderliness parameter* defined by the relation:

$$\Theta_{Yj} = Y_j W_{Yj} + C_Y \qquad (8.5)$$

where W_{Yj} is a weighting function measuring the deviation of Y_j of jth realization from those of other realizations of the state variable; and C_Y is a conditional coefficient which sets $d_0(\Theta_{Yj}) = 0$ when $Y_j = 0$. In writing the above relation, it is presumed that the system is *ergodic* with the entire ensemble of the parameter space having a common functional relation d_0.

Thus, the disorganization is an ensemble-generalized characteristic of disorder in the state of the neural system, weighted primarily by the probability of encountering a cell with a disorderly behavior and secondarily by its *relevance*. The relevance of the disorder to the jth situation is decided both by the functional relation d_0 common to the entire ensemble and the additional weight W_{Yj} accounting for the deviation of Y_j with the corresponding disorderliness of situations other than j.

The functional relation d_o pertinent to the self-organizing control endeavors of a neural complex refers to the sensitivity of the high-level goal towards the degree of failure to attain the subgoals considered. Such a failure or deviatory response arises due to the entropy of the system. Therefore, more explicitly O_D can be written as an entropy function in terms of Hartley's law as:

$$O_D = \sum_{j=1}^{N} p_j \log(\Theta_{Yj}) \tag{8.6}$$

Considering both temporal and spatial disorganizations associated with interconnected neurons, a superposition leads to the following general expression for the summation of the effects:

$$O_D^{\Sigma} = \sum_{\alpha=1}^{\ell} \tau_\alpha \sum_{\beta=1}^{m} r_\beta \sum_{j=1}^{N} p_j d_o(\Theta_{Yj}) \tag{8.7}$$

From the foregoing discussions it is evident that the disorganization of a neural complex is the consequence of:

- Spatial factors conceived in terms of the random locations of the neuronal cells.
- Temporal characteristics as decided by the random occurrence of state transitions across the interconnected cells.
- Stochastical attributes of the neural complexity.
- Combinatorial aspects due to the number and variety of the participating subsets in the neural activity.

Referring to spatiotemporal disorderliness as indicated earlier, let the *a priori* probability p_j denote the probability of occurrence (in time and space) of the jth neural event. When $p_j \rightarrow 1$, it amounts to a total disorderliness with $Y_j \rightarrow 0$. Likewise, $p_j \rightarrow 0$ sets a total disorderliness with $Y_j \rightarrow \infty$. The above conditions are met simultaneously by the following coupled relations:

$$Y_j = \left(1/p_j\right) - 1 \tag{8.8a}$$

$$\Theta_{Yj} = 1/p_j \tag{8.8b}$$

Therefore, the Hartley measure of the extent of disorganization can be written as:

$$O_D = \sum_{j=1}^{m} p_j \log\left(1/p_j\right) = -\sum_{j=1}^{m} p_j \log(p_j) \tag{8.9}$$

which is again in the standard form as *Shannon's statistical measure of entropy.*

The structural or the combinatorial aspect of disorganization pertains to the number of alternatives (such as the paths of state-transition proliferation or traces) in the interconnected network of cellular automata. Usually, only n out of such N alternatives are warranted to confirm a total orderliness. Therefore, the disorderliness is written as:

$$Y_j = (N_j/n) - 1 \tag{8.10}$$

so that $Y_j \to \infty$ for $N_j \to \infty$ with n being a constant and $Y_j \to 0$ for $N_j \to n$. The corresponding measure of disorganization written in conformity with Hartley's measure of entropy simplifies to:

$$O_D = \log(Y_j) \tag{8.11}$$

8.6 Entropy of Neurocybernetic Self-Regulation

In the self-organization endeavor, the neurocybernetic system attempts to (self) regulate (*via* feedback techniques), the state of the system (corresponding to say the ith realization of cellular state variable being nondeviatory with respect to a target state). That is, designating the generalized state of the system vector by y_i in the ith realization from the target vector y_T, the corresponding diversion is represented by $(y_i - y_T)$ and $|y_i - y_T| = Q_i$. A bounded two-dimensional spread-space, Ω_S includes all such likely diversions as depicted in Figure 8.2. It can be decomposed into elementary subspaces $\Delta\Omega$ which can be set equal to δ^2, where δ represents the one-dimensional quantizing level of the space.

Suppose the *a priori* probability (p_i) of finding the vector y_i at the ith elementary subspace is known, along with the distance of this i^{th} realization from the goal (target), namely, Q_i. The corresponding diversion ensemble of the entire spread-space $\{\Omega_S\}$ can be written as:

$$\{\Omega_S\} = \begin{bmatrix} |Q_1| & |Q_2| & ... & |Q_\kappa| \\ p_1 & p_2 & ... & p_\kappa \\ W_1 & W_2 & ... & W_\kappa \end{bmatrix} \tag{8.12}$$

where $\sum_{i=1}^{\kappa} p_i = 1$, and the number of form realizations in each subspace over the domain Ω_S is $\kappa = \Omega_S/\Delta\Omega$. Assuming equal weights, namely, $W_1 = W_2 = ... = W_\kappa = 1$, a goal-associated positional entropy $\mathcal{H}_y$ can be defined as follows:

$$\mathcal{H}_y = \sum_{i=1}^{\kappa} p_i \log[W_i\{ \mid y_i - y_T \mid - D(y_i) + 1\}] \tag{8.13}$$

The function $\mathcal{H}$ defines a new entropy space $(\Omega_{\mathcal{H}})$ as distinct from the parameter space Ω_S. That is, each value of $\mathcal{H}_y$ in the entropy space could be mapped onto the parameter spread-space and denoted on a one-to-one basis by $\mathcal{H}_y'$ $(\mid y_i - y_T \mid)$. This value of $\mathcal{H}_y'$ represents the expected mean diversion from the goal in the spread-space Ω_S. Further, the spaces Ω_S and $\Omega_{\mathcal{H}}$ are *affinely similar* and the goal-related entropy satisfies the conditions specified in the following lemmas [111]:

Lemma 1: $\mathcal{H}_y = 0$, if all $\mid y_i - y_T \mid \leq D(y_i)$ or if $p = 0$ for all
$\mid y_i - y_T \mid - D(y_i) > 0$ (8.14)

Lemma 2: $\mathcal{H}_y \to 0$ for the ensemble $p > 0$, if $\mid y_i - y_T \mid - D(y_i) \to 0$, and

$$\mathcal{H}_y \to \infty, \text{ if } \mid y_i - y_T \mid - D(y_i) \to \infty \tag{8.15}$$

Lemma 3: At $p = 1/\kappa$,

$$\mathcal{H}_y = -(1/\kappa) \sum_{i=1}^{\kappa} \log\{1/[\mid y_i - y_T \mid - D(y_i)]\} + \varepsilon_\kappa \tag{8.16}$$

where

$$\varepsilon_\kappa = (1/\kappa) \sum_{i=1}^{\kappa} \log\{[\mid y_i - y_T \mid - D(y_i)]/[\mid y_i - y_T \mid - D(y_i) + 1]\}$$

and

$$\varepsilon_\kappa \to 0, \text{ if } \mid y_i - y_T \mid - D(y_i) >> 1$$

Lemma 4: Sum of two entropies satisfying the conditions of independence and summation in the spread-space of the state vector leads to:

$$\mathcal{H}_{y(1,2)} = \log\{\mathcal{H}_{y_1}[\,|\,y_1 - y_T|\,] + \mathcal{H}_{y_2}[\,|\,y_2 - y_T|\,]\} + \varepsilon_{(1,2)}$$

$$(8.17)$$

where

$$\varepsilon_{(1,2)} = \log\{[\mathcal{H}_{y_1}(Q_1) + \mathcal{H}_{y_2}(Q_2)]/[H_{y_1}(Q_1) + \mathcal{H}_{y_2}(Q_2) + 1]\}$$

$$\to 0 \text{ for } \mathcal{H}_y(Q_i) \gg 1 \text{ with } Q_i = |\,y_i - y_T|, \quad (i = 1, 2)$$

Lemmas 1 and 2 represent the intuitive concept of neural disorganization in the state of being controlled towards the goal. If an ideal control is perceived, it strikes the well-ordered target domain in all realizations specified by $\mathcal{H}_y = 0$. Diversions from the ideality of the ensemble with an increasing or decreasing trend enable $\mathcal{H}_y$ to increase or decrease, respectively.

Lemma 3 stipulates that, in the event of equiprobable diversions, the relation between the spread-space of the state vector and the entropy space is logarithmic with an error $\varepsilon_\kappa \to 0$ for $|\,y_i - y_T| \gg 1$.

Entropy associated with target-seeking endeavor is not additive. That is, goal-associated entropies cannot be added or subtracted directly in the entropy space. However, these superposition operations can be performed in the spread-space of the state vector. After the necessary extent of such operations are executed in the spread-space, the consequent effects can be translated to the entropy space.

Lemma 4 specifies the rule of additivity in the spread space, pertaining to independent goal-associated position entropies with an accuracy set by $\varepsilon_{(1,2)} \to 0$ with $\mathcal{H}_y(Q_{1,2}) \gg 1$.

8.7 Subjective Neural Disorganization

Shown in Figure 8.3 is an arrangement depicting the ensemble of the state-vector realizations grouped with a center which does not coincide with the center of order or the goal-vector. That is, the goal center may lie outside the spread space. This happens when the neural system has self-organizing characteristics controlled *subjectively*. The subjectiveness refers to the neural intellect that has an information (or entropy) from the source, classifies the situations in it, predicts its behavior, and makes decisions leading the processed information to a desired functional attribute. While the ensemble of subjective realizations of the state-vectors tend to lie in a spread domain, the goal-dictated (objective) state-variables may normally cluster outside this region as shown in Figure 8.3.

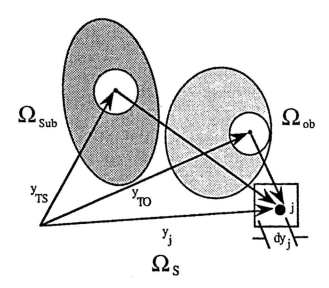

Figure 8.3 Objective and subjective neural disorganizations

The question of subjectiveness refers to the situation when the system prescribes inherently its own goal or subjective function regardless of the training-dictated objective function. This condition in the nervous system is, for example, due to external influences such as drug, alcohol etc. Such external factors may cause the control strategy of neural self-regulation to seek the subjective goal rather than the objective goal learned through experience.

Correspondingly, there are two possible disorganizations with respect to *subjective* and *objective goals* designated as y_{TS} and y_{TO}, respectively. The mutual positional disorganization defined between y_{TS} and y_{TO} is given by:

$$\mathcal{H}_{(o/s)y} = -\log[1/(g_{so} + 1)] \tag{8.18}$$

where g_{so} is the mutual diversion between the subjective and objective goals; and the corresponding *fractional position entropy* defined relative to the subjective goal is given by:

$$\Delta \mathcal{H}_{sy} = - \sum_{j=1}^{N} p_j \log[1/(\,|\,y_j - y_{TS}|\, + 1)] \tag{8.19}$$

where $|\,y_j - y_{TS}|$ is the spread about the subjective target. It should be noted that $\mathcal{H}_{(o/s)} \to 0$ as $g_{so} \to 0$, and the fractional positional entropy satisfies all the four conditions imposed earlier.

178

While y_{TO} is a steady-state (time-invariant) factor, y_{TS} may change subjectively with time. Therefore, $H_{(o/s)y}$ represents a dynamic parameter as decided by the dynamic wandering of the subjective goal center. In general, g_{so} is a stochastical quantity the extent of which is dictated by the neural (random) triggering induced by external influences. The concept of fractional positional entropy may find application in the analysis of a pathogenic neural complex and is a measure of *fault-tolerancy* in the network operation.

8.8 Continuous Neural Entropy

Reverting back to the situation wherein the subjective and objective targets merge, the spread space can be quantized into subspaces at the target center concentrically as shown in Figure 8.4 with a quantizing extent of spread space equal to δ.

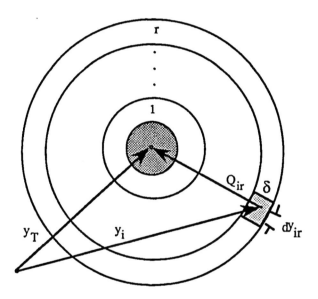

Figure 8.4 Discretized neural parametric spread-space

position entropy $\mathcal{H}_y$ by means of a nonlinear coefficient $\chi(\mathcal{H})$, such that $\mathcal{H}_y = \chi(\mathcal{H}) \mathcal{H}_{Sh}$, where $\mathcal{H}_{Sh}$ is the Shannon entropy which depends explicitly on the quantizing level δ as follows:

$$\mathcal{H}_{Sh} = - \int_{-\infty}^{+\infty} p(x)\log[p(x)\delta]dx \qquad (8.20)$$

Equation (8.20) predicts that in the limiting (continuous) case, $\mathcal{H}_{Sh} \to \infty$ if $\delta \to 0$ in confirmation with the fact that as the cardinality of the given set of events increases to infinity, the entropy of the system also inclines to follow a similar trend.

Suppose the vector y_i in the spread-space has an equal probability of occurrence at all subspaces Δy_{ir} measured from the target by the same distance $|y_{ir} - y_T| = Q_{ir}$. Then, for $\delta \to 0$, the probability of finding the vector y_{ir} in the rth annular space can be written as:

$$p_{Q_{ir}} = (2\pi Q_{ir}/\delta)p_{ir} \qquad (8.21)$$

satisfying the total probability condition that:

$$\int_0^{Q_{max}} p_Q \, dQ = 1 \qquad (8.22)$$

The corresponding position entropy is defined in a continuous form as:

$$\mathcal{H}_y = -\int_0^{Q_{max}} p_Q \log[(1/Q) + 1]dQ$$

$$= -\int_\infty^\infty p_Q \log[(1/Q) + 1]dQ \qquad (8.23)$$

with $0 \le Q \le Q_{max}$ and $Q = 0$ elsewhere.

The above expression is again identical to Shannon's entropy and represents the continuous version of Equation (8.13) with $\delta \to 0$. It can also be related functionally to Shannon's entropy $\mathcal{H}_{Sh}$ as before with $\mathcal{H}_{Sh}$ characterizing the statistical system disorganization for a constant set of cardinality. The goal-seeking position entropy $\mathcal{H}_y \to \mathcal{H}_{Sh}$ whenever $p_i \to 1/(Q_i + 1)$ for all realizations of the state vectors.

8.9 Differential Disorganization in the Neural Complex

The difference in entropies at two locations, namely, $\mathcal{H}_G$ in the vicinity of the objective locale of the goal and $\mathcal{H}_i$ at an arbitrary i^{th} location in the spread-space, measures implicitly the differential extent of disorganization. When $\mathcal{H}_i \to \mathcal{H}_G$, the corresponding *information gain* is given by $\Delta\mathcal{H} = (\mathcal{H}_i - \mathcal{H}_G)$.

The information or entropy pertinent to the disorganized i^{th} locale can be specified either by an associated *a priori* probability of attaining the goal

p_{ai} so that $\mathcal{H}_i = -\Sigma^k_{i=1}p_i \log(p_i/p_{ai})$, or in terms of conditional probability specified by p_i/p_{ai}' where p_{ai}' is the *a posteriori* probability of attaining the goal in reference to i^{th} subspace. The corresponding value of $\mathcal{H}_i'$ is equal to $-\Sigma^k_{i=1}p_i \log(p_i/p_{ai}')$. In both cases, p_i in general is the actual probability of the ith subspace and $p_i \neq p_{ai}$. Further, p_{ai}' depicts the forecast probability of the ith subspace. The value of differential measure of disorganization $\Delta\mathcal{H}$ is obtained by using $\mathcal{H}_i$ with $p_1 = p_2 = ... = p_K$ and the pragmatic aspects of $\Delta\mathcal{H}$ can be deduced from H_i'.

8.10 Dynamic Characteristics of Neural Informatics

The stochastical aspects of a neural complex are invariably dynamic rather than time-invariant. Due to the presence of intra- or extracellular disturbances, the associated information in the neural system may degrade with time; and proliferation of information across the network may also become obsolete or nonpragmatic due to the existence of synaptic delays or processing delays across the interconnections. That is, aging of neural information (or *degenerative negentropy*) leads to a devalued (or *value-weighted*) knowledge with a reduced utility. The degeneration of neural information can be depicted in a simple form by an exponential decay function, namely:

$$\mathcal{H}(t) = \mathcal{H}_0[1 - \exp(-t/\tau_{\mathcal{H}})] \tag{8.24}$$

where $\mathcal{H}_0$ is the initial information and $\tau_{\mathcal{H}}$ is the time constant specifying the duration within which $\mathcal{H}$ has the pragmatic value. This time constant would depend on the rate of flow of syntactic information, information-spread across the entropy space, the entropy of the sink which receives the information, and the characteristics of synaptic receptors which extract the information. Should a degradation in information occur, the network loses the pragmatic trait of the intelligence and may iterate the goal-seeking endeavor.

Another form of degradation perceived in neural information pertains to the delays encountered. Suppose the control-loop error information is delayed when it arrives at the controlling section; it is of no pragmatic value as it will not reflect the true neural output state because the global state of the neural complex would have changed considerably by then. In other words, the delayed neural information is rather devalued of its "usefullness" (or attains nonpragmatic value) at the receiving node, though the input and output contents of syntactic information remained the same.

In either case of information degradation, the value of information (devalued for different reasons) can be specified as $\lambda_{\mathcal{H}} = \lambda_{max} \exp(-t/\tau_{age})$ where λ depicts the pragmatic measure of information. It is also likely that there could be information enhancement due to redundancies being added in the information processor. This redundancy could be a predictor (or a precursor such as the synchronization of cellular state-transitional spikes) of information which would tend to obviate the synaptic delays involved. The corresponding time dependency of $\mathcal{H}$ can be represented as $\lambda_{\mathcal{H}} = [1 - \exp(t/\tau_{en})]\lambda_{max}$ where τ_{en} is the *enhancement time constant*.

The information aging and/or enhancement can occur when the neural dynamics goes through a nonaging or nonenhancement (quiescent) period. This quiescent period corresponds to the *refractory effects* in the neurocellular response.

The property of neural information dynamics can be described by appropriate *informational transfer functions*. Depicting the time-dependency of information function as $\mathcal{H}(t)$, its Laplace transform $\mathcal{H}(s)$ is the informational transfer function which describes the changes in the processor algorithm of the neural network.

The loss of neuronal information due to degradation can be specified by an *informational efficiency* which is defined as:

$$\eta_{\mathcal{H}_i} = \sum_{s=1}^{\kappa} [\mathcal{H}_{yi}/\mathcal{H}_{yi(max)}]_s \tag{8.25}$$

where $\mathcal{H}_{yi(max)}$ is the maximum usable information in the system and $\mathcal{H}_{yi}$ is the available information regarding the ith subset in the spread space. This informational efficiency is distributed among the sections of the neural network, namely, the input section, the output unit, and the control processor.

8.11 Jensen-Shannon Divergence Measure

As defined earlier, the disorderliness refers to a measure of deviation of a selected variable, say y_i, with reference to a specified target standard of order y_T. In a geometrical representation, y_T can be denoted by a vector corresponding to the center of a region of orderliness wherein a stipulated stochastical extent of orderliness is specified within bounds. The disorderliness at the ith realization in the parameter spread space is, therefore, given by Equation (8.2) with j replaced by i. That is:

$$Y_i = |y_i - y_T| - D(y_i) \tag{8.26}$$

where $|y_i - y_T| = Q_i$ refers to the magnitude of error vector and $D(y_i)$ is the distance from the center to the boundary of the quasi-ordered region. The corresponding goal-associated positional entropy is specified by $\mathcal{H}_{y_i}$ at the ith elementary subspace in the entropy space $\Omega_{\mathcal{H}}$. Elsewhere, say at j^{th}

subspace, let $\mathcal{H}_{y_j}$ represent the goal-associated positional vector entropy perceived. Now, one can seek information in the sample-space of y for discrimination in favor of $\mathcal{H}_{y_i}$ against $\mathcal{H}_{y_j}$, or symmetrically for discrimination in favor of $\mathcal{H}_{y_j}$ against $\mathcal{H}_{y_i}$. Physically, these conditions represent whether the ith realization or the jth realization (respectively) would enable achieving the goal being sought. The basic set of elements y_i's represent univariates and multivariates, discrete or continuous; it may simply represent the occurrence or nonoccurrence of an error signal of the control-processing loop. The elements have probability measures p_i's which are absolutely continuous with respect to each other. These probabilities can be specified by generalized probability densities $f(y_i)$ such that $0 < f(y_i) < \infty$ and $p_i = \int f(y_i) \, dy_i$ with $0 \le p_i \le 1$. The average information for discrimination in favor of $\mathcal{H}_{y_i}$ against $\mathcal{H}_{y_j}$ can be written as:

$$I(i: j, y) = (1/p_i) \int \log[f(y_i)/f(y_j)] dp_i \tag{8.27}$$

Considering the average information for discrimination in favor of $\mathcal{H}_{y_j}$ against $\mathcal{H}_{y_i}$, namely, I (j: i, y), a symmetric measure of divergence (known as the *Kullback measure*) can be written as [112,113]:

$$J(i, j) = I(i, j; y) + I(j, i: y)$$

$$= \sum_{y_i} (p_i - p_j)\log(p_i/p_j) \tag{8.28}$$

which is known as *J-divergence* and in the present case represents the divergence of disorganization associated with the subspace regions of ith realization and that of jth realization.

More appropriately, each of the realizations should be weighted in their probability distributions to specify their individual strength in the goal-seeking endeavor. Suppose Π_i and Π_j (Π_i, $\Pi_j \ge 0$ and $\Pi_i + \Pi_j = 1$) are the weights of the two probabilistic p_i and p_j, respectively. Then, a generalized divergence measure (known as the *Jensen-Shannon measure*) can be stipulated as follows [114]:

$$JS_\Pi (p_i, p_j) = \mathcal{H}(\Pi_i p_i + \Pi_j p_j) - \Pi_i \mathcal{H}(p_i) - \Pi_j \mathcal{H}(p_j) \tag{8.29}$$

This measure is nonnegative and equal to zero when $p_i = p_j$. It also provides upper and lower bounds for *Bayes' probability of error*. The Jensen-Shannon (JS) divergence is ideal to describe the variations between the subspaces or the goal-seeking realizations, and it measures the distances between the random graph depictions of such realizations pertinent to the entropy plane $\Omega_{\mathcal{H}}$.

In a more generalized form, the Jensen-Shannon divergence measure can be extended to the entire finite number of subspace realizations. Let p_1, p_2, ..., p_κ be κ probability distributions pertinent to κ subspaces with weights Π_i, Π_2, ..., Π_κ, respectively. Then the generalized Jensen-Shannon divergence can be defined as:

$$JS_\Pi (p_1, p_2, ..., p_\kappa) = \mathcal{H}\left[\sum_{i=1}^{\kappa} \Pi_i\, p_i\right] - \sum_{i=1}^{\kappa} \Pi_i\, \mathcal{H}(p_i) \qquad (8.30)$$

where $\Pi = (\Pi_i, \Pi_2, ..., \Pi_\kappa)$.

The control information processing unit sees κ classes $c_1, c_2, ..., c_\kappa$ with *a priori* probabilities $\Pi_1, \Pi_2, ..., \Pi_\kappa$. Each class specifies a distinct or unique strength of achieving the goal or minimization of the distance between its subspace and the objective goal. Now, the control information processing faces a decision problem pertinent to Bayes' error for κ - classes written as:

$$P(\varepsilon) = \sum_y p(y)[1 - \max\{p(c_1/x), p(c_2/x), ..., p(c_\kappa/x)\}] \qquad (8.31)$$

This error is limited by upper and lower bounds given by [114]:

$$(1/2)[\mathcal{H}(\Pi) - JS(p_1, p_2, ..., p_\kappa)] \leq P(\varepsilon) \geq \mathcal{H}_s /4(\kappa - 1) \qquad (8.32)$$

where $\mathcal{H}_s = \left[\mathcal{H}(\Pi) - JS(p_1, p_2, ..., p_\kappa)\right]^2$; and $\mathcal{H}(\Pi) = -\sum_{i=1}^{\kappa} \Pi_i\, \log\,(\Pi_i)$ and $p_i(y) = p(y/c_i)$, $i = 1, 2, ..., \kappa$.

The JS divergence measure can be regarded as an estimate of the extent of disorganization. Performance of the neural complex in the self-control endeavor to reduce the disorganization involved can be determined by maximizing an objective function criterion, subjected to certain constraints. This can be accomplished in an information domain *via* achieving a minimal JS- divergence. Denoting the information pertinent to the subsets y_i and y_j as $\mathcal{H}_i$ and $\mathcal{H}_j$, respectively, and the JS-divergence measure by $C_{JS}(\mathcal{H}_i, \mathcal{H}_j)$, a *feature vector* depicting the performance criteria of the control information process in the neural complex can be written as follows:

$$\mathcal{H}_j = \min\left[\sum_{i=1}^{\kappa} C_{JS}(\mathcal{H}_i, \mathcal{H}_j)\, p(\mathcal{H}_i/y)\right] \qquad (8.33)$$

where the JS measure $C_{JS}(\mathcal{H}_i, \mathcal{H}_j)$ characterizes the loss of organizational feature of the system as indicated above, while referring the subset y_j to subset y_i, and κ is the number of subsets being considered. Relevant to the performance criteria stipulated above, one should know the estimated conditional density function $p(y/\mathcal{H}_i)$ for all values of the vector y, which may however become formidable with a large number of subsets as in the case of the neural complex.

Therefore, a generalized control criterion can be specified with the arguments by groups, each group comprising of limited subsets $\{y_1, y_2, ..., y_m\}$. The corresponding performance criterion is hence given by:

$$\mathcal{H}_j = \min\left| C_{JS}(\mathcal{H}_j/P_i, P_2, ..., P_\ell)\right| \qquad (8.34)$$

where $P_1, P_2, ..., P_\ell$ are ℓ groups of subsets considered and:

$$C_{JS}(\mathcal{H}_j/P_1, P_2, ..., P_\ell) = \sum_{i=1}^{\kappa} C_{JS}(\mathcal{H}_i, \mathcal{H}_j)\, p(\mathcal{H}_j/P_1, P_2, ..., P_\ell) \qquad (8.35)$$

which refers to the divergence from the objective characterized by the *parsed* groups of subsets, namely, $(P_1, P_2,..., P_\ell)$ in reference to the entropy measure $\mathcal{H}_j$. The number of groups, each with a fixed number of subsets m, is determined by the combination $^m C_k$. Let the set $\{P_1, P_2, ..., P_\ell\}$ be denoted by $\{P_s\}_{s=1,2,...,\ell}$ and the subsets $(y_1, y_2, ..., y_\kappa)$ be divided into two sample spaces such that the first group $(y_1, y_2, ..., y_h)$ represents the subsets which are under the learning schedule to reach the objective function and $(y_{h+1}, y_{h+2}, ..., y_\kappa)$ are the subsets which have been annealed so that the dynamic states of the neurons have almost reached the steady-state condition. The (first) learning group can therefore be denoted by $\{P_L\}_{L=1,2,...,h}$ and the second group (which is closer to the objective function) is $\{P_M\}_{M=h+1,h+2,...,\kappa}$.

The differential measure of the features in these two groups can be specified by a matrix F represented as follows:

$$F_{L,V} = (1/h) \sum_{i=1}^{h} \mathcal{H}_j\Big(\{P_L\};\{P_V\}\Big)_V - \min\Big| C_{JS}(\mathcal{H}_j/P_L,P_V)_i\Big| = 0; \quad V < L$$

(8.36)

The corresponding average error $<\varepsilon>$ is evaluated as

$$<\varepsilon> = (1/q^h) \sum_{P_s} \sum_{i=1}^{h} \mathcal{H}_j(P_1, P_2, ..., P_\ell)_i - \Big| \min C_{JS}(\mathcal{H}_j/P_1, P_2, ..., P_\ell)_i \Big|$$

(8.37)

where q is the number of subsets in the group of converged subsets (P_M) and the summation is performed only in respect to subsets belonging to P_M. Further, $0 \le <\varepsilon> \le 1$, $s = 1, 2, ..., \ell$ and $(P_1, P_2, ..., P_\ell) \subset P_M$.

The back-propagation algorithm in the neural network attempts to gradually reduce the error specified by Equation (8.37) in steps. To increase the rate of algorithmic convergence, the divergence C_{JS} can be weighted at each step of the algorithm so that $\zeta = (1 - <\varepsilon>)$ approaches a maximum $\zeta_{max} \to 1$. This condition is represented by

$$\zeta_j = [1 - \sum_{P_s} \mathcal{H}_j(P_1, P_2, ..., P_\ell)_i | C_{JS}(\mathcal{H}_j/P_1, P_2, ..., P_\ell)_i |]/q_j^\kappa$$

(8.38)

where $s = 1, 2, ..., \ell$ and $(P_1, P_2, ..., P_\ell) \subset P_M$. Further, $q = \sum_{i=1}^{\kappa} q_i$ and the elements of the divergence matrix at the rth step of the algorithm are determined as follows:

$$[C_{JS}]_{ij}^r = [C_{JS}]_{ij}^{r-1} [\xi]_j^{r-1} \qquad i, j = 1, 2, ..., \kappa$$

(8.39)

The method of taking arguments into account by parsed groups constitutes an algorithm of evaluating a generalized objective criterion with initial information being insufficient to compute all the values of $p(y/\mathcal{H}_i)$ for a complete vector set **y**.

8.12 Semiotic Framework of Neuroinformatics

As discussed earlier, the neural complex bears three domains of informatics: The first one embodies the input, the second one is a processor stage, and the third part refers to the controlling stage.

The associated knowledge or the information in the first section is largely *descriptive* on the environment of the domain space accommodating the neurons. It represents a *declarative* knowledge which reflects the structure and composition of the neural environment. On the contrary, the second section has activities which are processed in translating the neuronal state-transitions across the interconnected cells as per certain instructions, rules, and set of learning protocols. In other words, a set of rule-based algorithms depicting the knowledge (or information) useful to achieve the goal constitutes the processing sections of the neural complex. The associative memory of a neural network immensely stores this declarative information.

Likewise, the controlling section of the network has information looped into the system to refine the achievement of the goal or minimize the organizational deficiency (or the divergence measure) which depicts the offsets in realizing the objective function. In this self-control endeavor, the neural automaton has the knowledge or information which is again largely *procedural*. It represents a collection of information which on the basis of phenomenological aspects of neurons reflects the rational relationship between them in evaluating the organizational deficiency.

Therefore, the neural complex and its feedback system harmoniously combine the declarative and procedural information. Should the control activity rely only on procedural informatics, it represents the conventional or classical model. However, due to the blend of declarative informatics, the neural controlling processor is essentially a *semiotic model* — a model that depicts a sum of the declarative knowledge pertinent to the controlled object, its inner structures, characteristic performance and response to control actions. The control actions themselves are, however, rule-based or procedural in the informatic domain.

The cohesive blend of declarative and procedural informatics in the neural automaton permits its representation by a semiotic model. That is, the neural complex is a system that could be studied *via semiotics* or the *science of sign systems*.

The semiotic modeling of the neural network relies essentially on the information pertinent to the controlled object, the knowledge on its control being present in the associated memory so that the system can be taught (or made to learn) and generate procedural knowledge by processing the stored control information [115].

To understand the applicability of semiotic concepts in neural activity, a fragment of interconnected network of neuronal cells is illustrated in Figure 8.5. The fragmented network has a set of cells {X, Y, Z, R, S, f, a, b, c} with axonal interconnections at synapses denoted by the set of numbers {1, 2, 3, 4, 5}. The synapses can be in either of the dichotomous states, namely, excited or inhibited. If some of the synapses are excited, the respective cell also becomes excited and this state is transmitted along all the outgoing axons which eventually terminate on other synapses. A synapse attains the excited state only when all the incoming axons are activated.

187

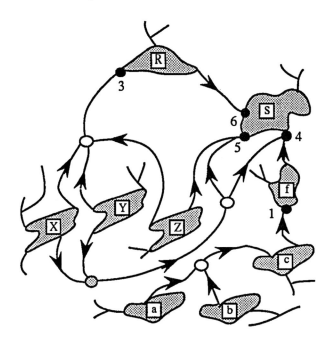

Figure 8.5 Semantic network model of the neural complex

The above network has three features: *syntax*, *semantics*, and *pragmatics*. The syntax refers to *signs* or symbols associated with the network. Syntactic representation breaks the network into fragments of cellular units or *parsed units*. Each syntax has a characteristic meaning referred to as the semantics. While a syntax has independent significance, the semantics of a sign can be meaningful only within the universal set of syntactics. The pragmatic aspect of the network refers to the *use* of a sign when the system is under a state of activity.

With the aforesaid semiotic attributions, the fragment of a neural network as depicted in Figure 8.5 is termed as a *semantic network*. When an excited state sets in a cell (due to inherent or exterior cellular disturbances), it proliferates across the neural complex to activate the other cells of the interconnected network. Shortly, the dynamic state-transitional process will become stabilized to freeze the network into a static excited state.

Referring to Figure 8.5, let S denote the realization of an excited state in a particular cell and refer to an objective goal. It is assumed that there are two rule-based state-transitional proliferations which could accomplish this, one *via* the set of cells denoted by {a, b, c, f} and the other through the set {X, Y, Z, R}. Considering the first possibility, let the cells {a, b, c} be initially activated. All the associated axonic outputs will then become activated. This will lead to the initiation of the synapse 1 at the cell f; and then, through the activation of this cell, an axonic output will render the cell S excited through

188

synaptic coupling, 4. The above procedure successfully meets the objective function or goal iff, S is excited. Otherwise, a feedback mechanism will let the process repeated on the basis of information gained by a learning procedure until the objective is met.

The process of activation to achieve the goal can be depicted in terms of semantic network representations with the associated procedural information pertinent to the network. Let every synapse stand for a certain procedure stored in the network memory (*via* training). The numerical signs (1, 2, ..., etc.) designate the name of the procedure, and the activation of the synapse is equivalent to the call of the respective procedure (from the memory) for execution. One can associate the initiation of the network cells with the presence of certain starting information required to solve the problem (or to attain the goal). In the present case, this information is designated by syntactics (a, b, c, ..., etc). The initiation of synapse 1 may call for a procedure of evaluating, say, func(c). Subsequently, synapse 4 calls for a procedure say, xy[func(c)]. If this end result is sufficient and successfully accomplished, the activation of S takes place and the goal is completed. Otherwise an error feedback would cause an alternative protocol (within the scope of learning rules). Nonconvergence to the objective function could be due to the nonobservance of the syntactic rules (caused by noise, etc.). In this respect, a neural network works essentially as a *recognizer set of syntactics*.

An alternative procedure of goal-seeking could be as follows: The initial activation of cells {X, Y, Z} will excite the synapse 3 *via* an axonic input. The procedural consideration at 3 could be say, $(X + Y + Z)R$. Then subsequently, this information is routed to synapse 6, where the procedure could be a comparison of $(X + Y + Z)R$ and S. If the difference is zero, the goal is reached. Otherwise, again feedback error information is generated to reiterate the procedure by means of a learning rule. That is, achieving the goal renders the transition of the network into statistically active states with the search for any additional procedural information being terminated.

The fragmented or *parsed network* constitutes a tree of *top-down* or *bottom-up parse* with the object-seeking protocol commencing at a set of synapses and culminating at the goal site. In these parsed tree representations, the error-information feedback and the subsequent control processing (or towards the minimization of the error to achieve the objective function) refer to an error-seeking scheme. Relevant strategy makes use of a set of tokens called *synchronization set* (or *tokens*). The tokens are identities of a set of manageable subsets (termed as *lexemes*) of cellular states obtained by parsing the cellular universal set.

The exemplification of the semantic network as above indicates the roles of procedural and declarative information in neurocybernetic endeavors. The first one refers to the stored information as the traditional memory and the second as the specially arranged model (parsed tree) represented by the synaptic interconnections.

The information associated with the semiotic model of a neural control processor can use three types of languages for knowledge representation, namely, *predicative, rational,* and *frame languages.* The predicative language employs an algorithmic notation (like a formula) to describe declarative information. Rational languages in a semantic network state the explicit relationships between the cells; that is, they represent the weighting functional aspects of interconnections.

The gist of these relationships represents the procedural knowledge of executing a functional characteristic at the synapse. For this purpose, the rational language that can be deployed is a set of *syntagmatic chains.* These chains can be constituted by elementary *syntagma.* For example, the weight and functional relation across a pair of interconnected synapses 2 and 1 in Figure 8.5 can be specified by an elementary stygma $(\alpha_2 \, \beta_c \, \alpha_1)$ where α_2 and α_1 are the synaptic elements interconnected by a relation, β_c. Between nodes 2 and 1 in Figure 8.5, this relation includes the state of the cell and the weights associated between 2 and 1. Proceeding further from synapse 1 to synapse 4, the corresponding network activity can be depicted by $((\alpha_2 \, \beta_c \, \alpha_1) \, \beta_f \, \alpha_4)$ which corresponds to the information pertinent to the activation, S. The entire relational language representation, as in the above example, has both declarative and procedural information. Thus in tree-modeling, information on the neural data structure, functioning, and control of the neural complex can be represented in a *language format* constituted by syntagmatic chains.

Another possible neural information representation is in *frame language* which could be again declarative or procedural. The basic description frame is declarative and embodies such information which cannot be reduced without losing the meaning of the process or events involved. On the other hand, there exists an *octant frame* which sheds off *redundancy* and provides minimal information for the neuronal control process. In a semiotic model, the frame language summarizes the languages of the relational types into multilevel hierarchical descriptions of the cellular complex. Such hierarchy is constituted by declarative information on the environment (descriptions of the physioanatomical aspects of the cells, axonic interconnections, etc.), and procedural aspects of the activities in the neural network (excitations and/or inhibitory biochemical responses, synaptic delays, etc.). It should be noted that this information is inherent in the associated memory of the network as beds and realizations mentioned earlier.

8.13 Informational Flow in the Neural Control Process

Illustrated in Figure 8.6 is the flow chart of neural activity in the informational domains.

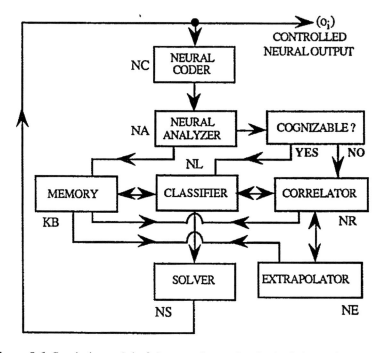

Figure 8.6 Semiotic model of the neural complex in the information domain
(Concept adapted from [115])

The functional aspects of the neural complex in the information plane as depicted by the semiotic model of Figure 8.6 can be summarized as follows: Pertinent to a set of input information that the neural complex receives from its environment, the information on the current output state of a neuron is fed back to the input of the network through an assembly of informational subsets. The first subset is a *neural encoder* (NC) which encodes the diverse information into syntagmatic chains. Such descriptions are fed to a *neural analyzer* (NA) which performs preliminary classification of the incoming information. If this information is needed for future use, it is stored in a *knowledge-base* (which is conventionally referred to as the memory (KB) of the neural complex) as traces. When the information in the neural analyzer is explicitly sufficient and cognizable, it is identified as information of the known category and directed to a *correlator* (NR) for the overlapping of the replicas, that is, for comparison with the trained patterns. Otherwise, it is considered as the information of unknown category and goes to a *classifier* (NL) for further categorization. The correlator serves to find a chain of control actions (to achieve the objective function) to match the controlling required. It also turns to the classifier to know the category of the information of the current neuronal state, if required. It may further use the data (memory) stored in the knowledge-base, as well. If the correlator finds a single control action, it informs the *solver* (NS) to let the relevant controlling activity to take place. If there are alternatives possible on the control strategy, the correlator may find it from the *extrapolator* (NE). The extrapolator seeks information from the knowledge-base pertinent to the output neuronal state and predicts the likely consequences of different control strategies. The responsibility of final controlling action rests with the solver.

Thus, semiotic modeling of neural control processing permits the development of the entire control protocols in the information plane; and optimization of this semiotic control process provides a scope for the developments in neuroinformatics.

Further, the semiotic approach in implementing the control information establishes its identity with the information existing at the input to the control. Thus, after being processed, the input information is transformed into control information which in turn provides input information about the neural state variables at the processor and, after the transit through the control means and the controlled system, input information about the functioning of the entire neural network. As dictated by the goal, each of the above stages is an information processing endeavor depicting sensing, recognition, prediction, decision making, and execution protocols. The pragmatics of such information and the characteristics of the processor can be evaluated as follows.

Considering the role of information processing in the control of neural disorganization, the useful information (pragmatic assay) in reducing the disorganization in the sensing subgoal is given by:

$$\mathcal{H}_s = \Delta O_{D(S)} = d_o[d_o^{-1}(\mathcal{H}_{in}) \mathcal{P}_v L_s \eta_S] \tag{8.40}$$

where $\mathcal{H}_{in}$ and $\mathcal{P}_v$ are the amount of syntactic input information and its pragmatic values, respectively, with regard to the sensing algorithm. L_s is the semantic processing characteristics of the sensing algorithm, η_s is the informational efficiency of the receiving synapse, and d_o^{-1} is the inverse of the functional relation specified by Equation (8.4).

The sensing subgoal refers to extracting only useful parts from the total information sufficient to impart knowledge to the controlled system. Similar to the above expression, the extent of useful information at the receiving synapse (recognizing the knowledge) can be written as:

$$\mathcal{H}_R = \Delta O_{D(R)} = d_o[d_o^{-1}(\mathcal{H}_s) \alpha_{R/S} L_s \eta_R] \tag{8.41}$$

where $\Delta O_{D(R)}$ refers to the reduction in disorganization with respect to recognition subgoal, $\mathcal{H}_s$ is the sensing information, $\alpha_{R/S}$ is the coefficient of coupling between the pragmatic value of sensing information and recognition subgoal, (semantic properties of recognition), and η_R is the associated informational efficiency.

Similar expressions can be written for prediction and decision-making information protocols aimed at minimizing the overall disorganization in the goal-related hierarchy.

8.14 Dynamic State of Neural Organization

As indicated in Section 8.10, the neuronal informational flow has distinct dynamic characterizations dictated by synaptic delays and cellular disturbances. The consequence of this is a devaluation of the pragmatic values of the information content. The dynamic state of informational flow portrays the dynamic degradation of self-regulating performance of the neural complex assessed *via* disorganization parameters in terms of the devalued information parameters.

The neural complex is essentially a self-regulating automaton which maintains its stability without assistance from the control system. Opening the control loop in such systems causes goal-related disorganization to increase within, however, certain bounds.

Intra- and/or extracellular disturbances cause the vector y in the spread space Ω_S to deviate from the system objective. Such diversions in the information plane refer to the *spatial transient diversions*. On opening the control-loop in a goal-seeking self-controlling system, the corresponding disorganization O_D^* can be described symbolically as:

$$O_D^* \rightarrow O_D^{int} \cup O_D^{ext}$$

$$= d_o[d_o^{-1}O_D^{int}] + d_o^{-1}O_D^{ext}) - d_o^{-1}(R_o) + C_Y \qquad (8.42)$$

where R_o is a subset of the quasi-ordered region of Ω_S and represents the well-ordered region or the region at which the goal is totally attained in the spread space, and C_Y is the constant as defined in Equation (8.5).

When the control-loop is closed, there is a net steady-state control information $(\mathcal{H}_c)$ *vis-a-vis* disorganization O_D. They are related by:

$$O_D = d_o[d_o^{-1}O_D^*] - d_o^{-1}(\mathcal{H}_c) + C_Y \qquad (8.43)$$

To let $O_D \rightarrow 0$, O_D^* should converge to $\mathcal{H}_c$, the control information. That is, an information invariance occurs when actions of the information processor and implicit organizational activities balance against each other so that the vector y converges to the objective locale. By the same token, there may be information invariance in each stage of information processing. Thus,

193

information invariance is a relative indicator of the extent of the objective been achieved.

Considering the dynamic state of the neural system, the control information $\mathcal{H}_c(t)$ and the corresponding state of disorganization $O_D(t)$, the transient (time-varying) information state of the system can be written as:

$$O_D(t) = d_o \{ d_o^{-1}[O_D^{int}(t)] + d_o^{-1}[O_D^{ext}(t)] - d_o^{-1}[R_o(t)] - d_o^{-1}[\mathcal{H}_c(t)] + C_Y \}$$

(8.44)

The conditions for the invariancy of this disorganization control information under the dynamic state which stipulate the convergence towards the objective function cannot, however, be ascertained explicitly without approximations.

8.15 Concluding Remarks

The parallel theme of cybernetics and informatics as posed in this chapter is a coordinated approach in seeking a set of avenues to model the information processing in the neural complex *via* entropy of disorganization and semiotic considerations. In this endeavor, it is strived to provide relevant conceptual and mathematical foundations of the physical process involved supplemented by the host of definitions and terminologies of various quantifiable terms which dictate the neural functions in the informatic plane. The concepts of cybernetics are matched with functioning of the neural complex through the general principles of information theory. This admixture of analysis could possibly lead to the passage of viewing higher order, composite neural networks (interacting networks) which (unlike the simple interacting networks) are more involved at organizational and operational levels calling for new strategies of neural information handling.

The informatic structure of neural assembly is complex. It has hierarchical relationships between and within the control centers. The neural channels, in general, handle huge informational flows. There is an explicit dependence of neural information and the attainment of the objective function. In fact, the processing involved in the realization of the target (objective function) specifies the sematic and pragmatic aspects of the data proliferating across the network of neural cells. The neural information base with its hierarchical structure of control and goal has a system-wide domain enclosing memory locales and control centers with distinct functions warranting specific algorithmic executions. The relevant characteristics of information handled pertain to disorganization in the (mal)functioning of the neural complex with regard to its objective function. Such a disorganization, in general, is a dynamic (time-dependent) entity.

A neurocybernetic system attempts to self-regulate despite of it being informatically starved. That is, more often a neural complex is (self) controlled under conditions of incomplete information. This is due to the fact that a massive system like the neural assembly is only partly observable and controllable (by the C^3I protocol), and consequently partly cognizable and predictable towards self-regulation (or in achieving the objective function). The variety features dictated by the extremely complex structure, composition and properties of the neural units and the limited time of operation involved (or available) in the control endeavor are responsible for the incomplete acquisition of the neural information. Nevertheless, analysis of the neurocybernetic system in the informatic plane permits a singular entity (namely, the entropy) to depict the functioning of the neural complex system (both in terms of memory considerations as well as control information processing) and provides a framework for alternative analytical strategies to study the neural complex.

APPENDIX A

Magnetism and the Ising Spin-Glass Model

A.1   General

Materials in which a state of magnetism can be induced are called magnetic materials. When magnetized, such materials cause a magnetic field (force) in the surrounding space.

The molecules of a material contain electrons which orbit around their nuclei and spin about their axes. Electronic motion equivalently represents an electric current with an associated magnetomotive force. That is, a magnetic material consists of a set of atomic level magnets arranged on a regular lattice which represents the crystalline structure of the material. In most molecules each magnetomotive force is neutralized by an opposite one, rendering the material nonmagnetic. However, in certain materials (such as iron), there is a resultant unneutralized magnetomotive force which constitutes a magnetic dipole. These dipoles are characterized with dipole moments due to three angular momenta, namely, orbital angular momentum, electron-spin angular momentum, and nuclear-spin angular momentum. The atomic magnets represent the associated $\pm 1/2$ *spin* atoms in which the angular spin is restricted to two distinct directions and represented in the so-called Ising model by $S_i = \pm 1$ at a lattice site i.

In a magnetic material, the dipoles line up parallel with an externally applied magnetization field h^{ext} and internal field produced by other spins. When the dipoles remain aligned despite of removing external magnetization, the material is said to have acquired *permanent magnetism*. The readiness of a magnetic material to accept magnetism is termed as its *permeability (μ_M)*.

In the process of magnetization, the dipole moment experienced per unit volume quantifies the extent of magnetization (of the medium) and is denoted by M_M. The manner in which magnetic dipoles are arranged classifies the materials as *paramagnetic, ferromagnetic, antiferromagnetic,* and *ferrimagnetic*, as illustrated in Figure A.1.

With reference to the arrangements of Figure A.1, the nature of dipole interactions can be described qualitatively as follows:

Paramagnetic $\Rightarrow$ Disordered interaction with zero resultant magnetization.

Ferromagnetic $\Rightarrow$ Ordered strong interaction with a large resultant magnetization.

196

Antiferromagnetic ⇒ Ordered interaction, but with zero resultant magnetization.

Ferrimagnetic ⇒ Ordered interaction, with relatively weak resultant magnetization.

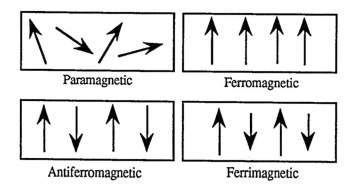

Figure A.1 Magnetic spin arrangements

A.2 Paramagnetism

Paramagnetism is a noncooperative magnetism arising from spontaneous moments, the orientation of which is randomized by the thermal energy k_BT. The magnetic order is thus the net alignment achieved by the applied field in the face of the thermal disarray existing at the ambient temperature. The extent of magnetization is the time-average alignment of the moments, inasmuch as the orientations fluctuate constantly and are in no sense fixed. Ideal paramagnetism is characterized by a susceptibility which varies inversely proportional to temperature *(Curie law)* and the constant of proportionality is called the *Curie constant* .

At any given lattice location, the internal field pertinent to an atomic magnet due to the other (neighboring) spins is proportional to its own spin. Thus, the total magnetic field at the site i is given by:

$$h_i = \sum_i J_{ij} S_i S_j + h^{ext} \tag{A.1}$$

where the summation refers to the contribution from all the neighboring atoms and the coefficient J_{ij} measures the strengths of influence of spin S_j on the field at S_i and is known as *exchange interaction strengths.*

(Another noncooperative magnetism known as diamagnetism is characterized by a negative temperature-dependent magnetic susceptibility.)

A.3 Ferromagnetism

As indicated earlier, ferromagnetism results from an ordered interaction. It is a cooperative magnetism with a long-range collinear alignment of all the moments, manifesting as the prevalence of magnetization even with no external magnetic field *(spontaneous magnetization)*. This cooperative magnetism becomes prominent and dominates below a discrete temperature called *Curie temperature* (T_C), and in spin-glass theory it is known as *spin-glass temperature* (T_G).

A.4 Antiferromagnetism

Like ferromagnetism, this is also a cooperative process with a long-range order and spontaneous moments. The exchange coupling dominates the constraints on the moments in the cooperative process, and the exchange parameter for an antiferromagnetism is negative. That is, for a given direction of the moment of an ion, there is a neighboring ion with its moment pointing exactly in the opposite direction. Hence, there is no overall spontaneous magnetization. The transition temperature in this case is known as *Néel temperature* (T_N).

A.5 Ferrimagnetism

This magnetism differs from the other types, as it involves two or more magnetic species that are chemically different. The species can have ferro- or antiferromagnetic alignment; however, if the species with ferromagnetic moment dominate there would be a resultant spontaneous magnetization. The temperature dependence of ferrimagnetism is qualitatively similar to that of ferromagnetism.

A.6 Magnetization

As described earlier, a permanently magnetized sample say, of iron, typically contains a number of domains with the directions of magnetization being different in neighboring domains. The extent of magnetization M_S that exists in the sample in the absence of any applied magnetic field is called *spontaneous magnetization* (more commonly known as *retentivity* or *remnant magnetism* in engineering). Its magnitude depends on the temperature in the manner sketched in Figure A.2.

The direction in which the spontaneous magnetization is rendered usually lies along one of several *easy axes* defined by the crystal structure of the material. Upon heating to the critical or Curie temperature T_C, the spontaneous magnetism vanishes continuously. The immediate neighborhood of T_C is called the *critical region* and $M_S \simeq (T_C - T)^{\beta_C}$ where β_C varies rather little from one ferromagnetic material to another and is typically about 1/3.

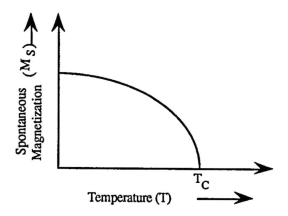

Figure A.2 Spontaneous magnetization *versus* temperature
T_C: Curie point

The property of the material in the neighborhood of the critical point is known as *critical phenomenon* which is substantially independent of the detailed microscopic constitution of the system considered.

The forces which tend to align the spins in a ferromagnet have electrostatic and quantum mechanical origins. The *phase transition* which occurs at the Curie temperature in zero magnetic field can be described as an *order-disorder transition*. In the high temperature (paramagnetic) phase, one in which the fluctuations in the orientation of each spin variable are entirely random so that the configurational average is zero, the state is a *disordered* one. In the ferromagnetic phase, on the other hand, inasmuch as the spins point preferably in one particular direction, the state is said to be *ordered*.

The interaction of the spins in a magnetic system dictates implicitly the minimum energy considerations which can be mathematically specified by a Hamiltonian. That is, considering the potential energy corresponding to Equation (A.1), it can be specified by a Hamiltonian H_M given by:

$$H_M = (-1/2) \sum_{ij} J_{ij} S_i S_j - h^{ext} \sum_i S_i \qquad (A.2)$$

A.7 Spin-Glass Model [116]

The phenomenon of ferromagnetism was modeled by Ising in 1925 [35] as due to the interaction between the spins of certain electrons in the atom making up a crystal. As mentioned earlier, each particle of the crystal is associated with a spin coordinate S. Assuming the particles are rigidly fixed in the lattice and either neglecting the vibrations of the cyrstal or assuming that they act independently of the spin configuration (so that they can be treated separately), the spin is modeled as a scalar quantity (instead of a vector) which can assume two dichotomous levels, namely, S = ±1 (+1 refers to the *up* spin and –1 is the *down* spin). The interaction between two

199

particles located at the ith and j^{th} lattice points is postulated as $E_{ij} = -J_{ij}S_iS_j$ if i and j are nearest-neighbors; otherwise $E_{ij} = 0$. This postulation assumes only nearest-neighbor interaction. That is, the energy is -J if the nearest neighbors have the same spin and +J if they have unlike spins; and the zero energy corresponds to the average of these two. Thus the parameter J is a measure of the strength of i to j coupling or the exchange parameter determined by the physical properties of the system. It is positive for a ferromagnetic system and negative for an antiferromagnetic system.

In the Ising model, it is further postulated that the particle can interact with an external magnetic field as well. Denoting the magnetic moment m_M assigned to a lattice point for an external magnetic field h_M applied, the associated energy of interaction of the ith particle can be written as $E_M = -m_M h_M S_i$. The corresponding thermodynamics of the system with M lattice points can be specified by a *partition function* * as:

$$Z = \sum_{S_i=\pm 1} \sum_{S_M=\pm 1} \exp\{(J/2k_BT)\sum_{ij=1}^{M} a_{ij}S_iS_j + (m_Mh_M/k_BT)\sum_{i=1}^{M} S_i\}$$

(A.3)

where $a_{ij} = 1$ if i and j are nearest neighbors, otherwise $a_{ij} = 0$, and further k_B is the Boltzmann constant and T is the temperature.

In terms of this partition function, the internal energy E_{int} per particle and the magnetization M_M per particle can be written as:

$$E_{int} = (M^{-1})(k_BT^2)\partial(\log Z)/\partial T$$

(A.4)

and

$$M_M = (M^{-1})\partial(\log Z)/\partial h_M^*$$

(A.5)

where

$$h_M^* = h_M/k_BT$$

(A.6)

The Ising partition function indicated above quantifies the scalar spin interaction which does not depend on the spin orientation with the lattice distance being fixed.

Although the Ising model is not considered as a very realistic model for ferromagnetism, it is considered as a good emulation of a binary substitutional alloy as well as an interesting model of a gas or liquid.

---

* The definition and details on *partition function* are presented in Section.A.13.

A.8 The Hamiltonian

The simplest model for a d-dimensional magnetic system is the Heisenberg Hamiltonian (in the absence of external field) given by:

$$H_M = -\Sigma J_{ij} S_i S_j \tag{A.7}$$

where S_i, S_j, etc. are the m-component spin vectors. When m = 1, the corresponding model is the Ising model. These vectors at the positions r_i and r_j, etc. (in the d-dimensional space) interact with each other, the strength of such interaction(s) being J_{ij} as mentioned before. In a general case, J_{ij} depends on r_i, r_j, etc. The exchange interaction J_{ij} on the right-hand side of Equation (A.7) is called *short-ranged* or *long-ranged interaction* depending on whether $\Sigma_j |J(r_{ij})| < \infty$ or $\Sigma_j |J(r_{ij})| = \infty$.

Commensurate with spin theory, the spontaneous magnetization is a measure of the long-range orientational (spin) order in space. It is an order parameter that distinguishes the ferromagnetic phase from the paramagnetic phase or the so-called *phase transition* .

A.9 Bonds and Sites

In a magnetic system controlled by the spins, the "disorder" discussed earlier can be introduced either at the "bonds" or at the "sites". The bond-random model assumes the exchange bond strengths as independent random variables with a specific probability density function (pdf). For example, when the exchange interaction is such that each spin is assumed to interact with every other spin in the system (specified as the lattice coordination number tending to infinity), the corresponding pdf of J_{ij} is shown to be gaussian.

In the site-disorder models, the randomness is due to a finite (nonzero) fraction of sites being occupied (randomly) by spins and the remaining sites being occupied by nonmagnetic atoms/molecules. The state of the disorder arises due to the finite temperature. That is, thermal fluctuations tend to flip the spins from down to up or from up to down, and thus upset the tendency of each spin to align with its field. At the so-called absolute zero temperature (-273°C or 0°K), such thermal fluctuations vanish. Conventionally, the effect of thermal fluctuations in an Ising model is depicted by *Glauber dynamics* with a stochastic rule:

$$S_i = \begin{cases} +1 \text{ with pdf } & p(h_i) \\ -1 \text{ with pdf } & 1 - p(h_i) \end{cases} \tag{A.8}$$

where the pdf depends on the temperature (T) with difference functional choices. Usually the following sigmoidal (S-shaped) function is considered for $p(h_i)$:

$$p(h_i) = 1/[1 + \exp(-2\beta h_i)] \qquad (A.9a)$$

where $\beta = 1/k_B T$ with $k_B = 1.38 \times 10^{-16}$ erg/K and is known as the Boltzmann constant. The dynamic rule of symmetry for the state S_i can be written as:

$$\text{prob}(S_i = \pm 1) = 1/[1 + \exp(\mp 2\beta h_i)] \qquad (A.9b)$$

The factor β (or the temperature T) controls the steepness of the sigmoid near $h = 0$ as shown in Figure A.3.

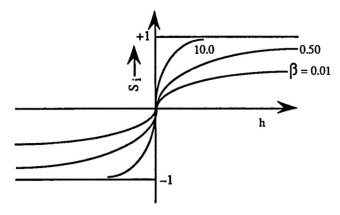

Figure A.3 Glauber's rule of symmetry for the state S_i

A.10 The Hard Spin

The Ising spin is referred to as *hard*, as the spins can have only fixed (finite) values, namely +1 and −1. In other words, the spins are characterized by a *weighting function* specified by:

$$\exp[-J_{ij}(S_i)] = \delta(S_i+1) + \delta(S_i - 1) \qquad \text{for all i} \qquad (A.10)$$

A.11 Quenching and Annealing

The disorder in a magnetic spin system can be classified as either *quenched* or *annealed*. Such a classification arises from the spatial ergodic property, namely, the physical properties of a macroscopically large system are identical with the same property averaged over all possible spatial configurations. Suppose a large sample is divided into a large number of smaller subunits (each of which is statistically large) in such a way that each of these subunits can be regarded as an independent number of ensemble systems characterized by a random distribution; and, as each of the subunits is statistically large, surface interaction can be assumed as negligible.

If the original system is *annealed*, it refers to each of the subunits acquiring all possible configurations in them because positions of the *impurities* are not frozen. Suppose the original system is *quenched*. Neglecting surface interactions, there is a frozen rigidity meaning that the positions of the impurities are clamped down. The result is that the characteristics of the quenched systems could only be the superposition of the corresponding characteristics of the subunits. The spin-glass systems have the common features of the quenched disorder.

A.12 Frustration

The interactions between spins are normally in conflict with each other leading to *frustration*. For example, the spin arrangement in a unit of square lattice (called a *plaquette*) can be of two possibilities as illustrated in Figure A.4.

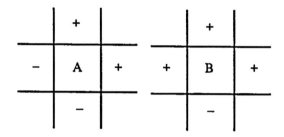

Figure A.4 Plaquettes of a spin system
A: Unfrustrated interaction; B: Frustrated interaction;
+, –: Two opposing spins

The individual bound energies are minimized if the two spins connected by an arbitrary bond <ij> are parallel to each other for + sign and antiparallel to each other for – sign. In plaquette A, all the bond energies can be minimized simultaneously whereas the same is not possible in plaquette B. Hence, plaquette B is called *frustrated*. In other words, those plaquettes where topological constraints prevent the neighboring spins from adopting a configuration with bond energy minimized are called frustrated. The extent of frustration is quantified by a *frustration function* dictated by the product of bond strengths taken over a closed contour of connected bonds.

A.13 Partition Function

In a physical system with a set of ν states, each of which has an energy level ϕ_ν, at a finite temperature $(T > 0)$ the fluctuation of states attains a thermal equilibrium around a constant value. Under this condition each of the possible states ν occurs with a probability $P_\nu = (1/Z)\exp(-\phi_\nu/k_BT)$ where Z is a normalization factor equal to $\Sigma_\nu \exp(-\phi_\nu/k_BT)$. That is associated with discrete states $\phi_\nu (\nu = 1,2,...)$ (each of which occurs with a probability under thermal equilibrium), a controlling function which determines the *average energy* known as the *sum of the states* or the *partition function* is defined by:

$$Z = \sum_\nu \exp(-\beta\phi_\nu) \tag{A.11}$$

In the presence of an applied magnetic field $(m_M h_M)$, the partition function corresponding to the Ising model can be specified *via* Equations (A.9) and (A.11) as:

$$Z^* = \sum_{S_i = \pm 1} \exp[\beta J \sum_{<ik>} S_i S_k + \beta m_M h_M \sum_i S_i] \tag{A.12}$$

where Σ_i is taken over all spins and $\Sigma_{<ik>}$ is taken over all pairs of direct neighbors. Further, $\Sigma_{S_i=\pm 1}$ is over the 2^M combinations of the M spins. The associated energy E and magnetic moment m_M can be specified in terms of Z^* as:

$$E = -[\partial \ln Z^*/\partial \beta]_{h_M} \tag{A.13}$$

$$m_M = (1/\beta)[\partial \ln(Z^*)/\partial h_M]_\beta \tag{A.14}$$

The relations specified by Equations (A.12), (A.13), and (A.14) are explicitly stated in the thermodynamics viewpoint with $\beta = k_BT$ in Equations (A.3), (A.4), and (A.5), respectively.

A.14 Ising Model: A Summary

Let a string of M identical units, numbered as 1, 2, 3, ..., (M−1), M each identified with a state variable x_1, x_2, x_3, ..., x_M represent a one-dimensional interacting system. A postulation of nearest-neighbor interaction is assumed which specifies that each unit interacts with its two direct neighbors only (see Figure. A.5).

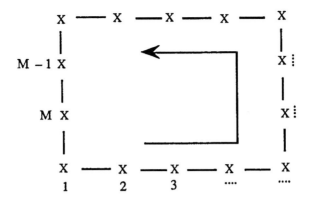

Figure A.5 A string of M cooperative interacting units

Let the interaction between two neighbors be $\phi(x, y)$. The corresponding probability for a given state of the system is proportional to the Boltzmann potential, namely, $\exp[-\beta\{\phi(x_1, x_2) + \phi(x_2, x_3) + ... + \phi(x_M, x_1)\}]$ so that the corresponding partition function can be written as a summing (or integration) function as defined in Equation (A.11). That is:

$$Z = \sum_1 \sum_2 ... \sum_M \exp[-\beta\{\phi(x_1, x_2) + \phi(x_2, x_3) + ... + \phi(x_M, x_1)\}]$$

(A.15)

On the basis of the reasoning from probability calculus, the following eigenvalue problem can be associated with the summing function:

$$\sum_y \exp[-\beta\{\phi(y, z)\}]a(y) = \lambda a(z)$$ (A.16)

where λ has a number of different eigenvalues λ_v, to each of which there belongs one eigenvector a_v. Also, an orthogonality relation for the a's prevails, given by:

$$\sum_y a_\mu(y)a_v(y) = \delta_{\mu v}$$ (A.17)

Hence, the following integral equation can be specified:

$$\exp[-\beta\{\phi(y, z)\}] = \sum_v \lambda_v a_v(y)a_v(z)$$ (A.18)

Using the above relations, Z reduces to:

$$Z = \sum_v \lambda_v^M \tag{A.19}$$

and

$$Z \approx \lambda_1^M \tag{A.20}$$

where λ_1 is the largest eigenvalue.

In the so-called one-dimensional Ising model as stated earlier, the variable x is the spin S which is restricted to just the two dichotomous values ± 1. The interaction $\phi(x, y)$ is therefore a matrix. In the case of a linear array of spins forming a closed loop (Figure A.5), the interaction $\phi(x, y)$ simplifies to:

$$\phi = [(-JSS') - m_M h_M (S + S')/2] \tag{A.21}$$

where J refers to the exchange coupling coefficient.

The Ising Hamiltonian with $h_M = 0$ has a symmetry; that is, if the sign of every spin is reversed, it remains unchanged. Therefore, for each configuration in which a given spin S_i has the value +1, there is another one obtained by reversing all the spins, in which case it has the value −1; and both configurations have, however, the same statistical weight. The magnetization per spin is therefore zero, which is appparently valid for any temperature and for any *finite* system.

Thus, within the theoretical framework of the Ising model, the only way to obtain a nonzero, spontaneous magnetization is to consider an *infinite* system, that is, to take the *thermodynamic limit* (see Figure A.6).

Considering single spins which can be flipped back and forth randomly between the dichotomous values ± 1 in a fixed external magnetic field $h = h^{ext}$, the *average magnetization* refers to average of S given by $<S> = prob(+1).(+1) + prob(-1).(-1)$ which reduces to $\tanh(\Lambda h)$ *via* Equation (A.9a).

Further considering many-spins, the fluctuating values of h_i at different sites can be represented by a mean value $<h_i> = \sum_j J_{ij} <S_i> + h^{ext}$. That is, the overall scenario of fluctuating many-spins is focused into a single average background field. This *mean field* approximation becomes exact in the limit of *infinite range interactions*, where each spin interacts with all the others so that the principle of *central limit theorem* comes into vogue.

A.15 Total Energy

Considering the Ising model, the state of the system is defined by a configuration of + (up) and − (down) spins at the vertices of a square lattice in the plane.

Each edge of the lattice in the Ising model is considered as an interaction and contributes an energy $\phi(S, S')$ to the total energy where S, S' are the spins at the ends of the edge. Thus in the Ising model, the total energy of a state σ is:

$$\phi(S, S') = \sum_{ij} k_1 S_{i,j} S_{i+1,j} + k_2 S_{i,j} S_{i,j+1} \tag{A.22}$$

The corresponding partition function is then defined by (as indicated earlier):

$$Z = \sum_{S} \exp[\phi(S)/k_B T] \tag{A.23}$$

A.16 Sigmoidal Function

The spin-glass magnetization (Figure A.6) exhibits a distinct S-shaped transition to a state of higher magnetization. This S-shaped function as indicated earlier is referred to as the *sigmoidal function*. In the thermodynamic limiting case for an infinite system at $T < T_C$, the sigmoidal function tends to be a step-function as illustrated. In the limit, the value of M_M at $h = 0$ is not well defined, but the limit of M_M as $h \to 0$ from above or below the value zero is $\pm M_S(T)$.

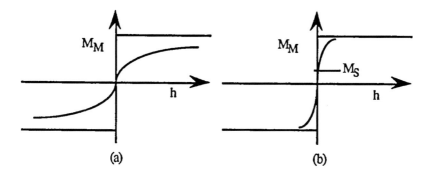

(a)   (b)

Figure A.6 Magnetization of the one-dimensional Ising chain as a function of the magnetic field
(a) Finite system $(T > T_c)$; (b) Infinite system $(T < T_c)$

(The method of calculating the actual values of magnetization under thermodynamic considerations is dimension-dependent. For the one-dimensional case, there is no ferromagnetic state. For the two-dimensional case, the solutions at zero magnetic field due to Onsager and for a nonzero field by Yang are available. For three dimensions, no exact solution has been found; however, there are indications of a possible ferromagnetic state.)

A.17 Free Energy

Pertinent to N atoms, the partition function is summed over all possible states, all the 2^N combinations of the spins, $S_i = \pm 1$. That is:

$$Z = \sum_{S_1 = \pm 1} \cdots \sum_{S_N = \pm 1} \exp[\beta/2 \sum_{ij} J_{ij} S_i S_j + \Gamma \sum h_i S_i] \tag{A.24}$$

The multiple summation over all states is known as the *trace* T^Σ. Hence the average activation $<S_i>$ of limit i is given by:

$$<S_i> = (1/Z)T^\Sigma \{S_i \exp[(\beta/2) \sum J_{ij} S_i S_j + \beta \sum_i h_i S_i]\} \tag{A.25}$$

For $Z = Z(h_i)$:

$$<S_i> = (1/\beta Z)\partial Z/\partial h_i = (1/\beta)\partial/\partial h_i[\log(Z)] \tag{A.26}$$

By defining a *free-energy* term as $F_E = -(k_B T)\log(Z)$, Equation (A.26) reduces to:

$$<S_i> = -\partial F_E/\partial h_i \tag{A.27}$$

and the correlation function $<S_i S_j>$ becomes:

$$<S_i S_j> = -\partial F_E/\partial J_{ij} \tag{A.28}$$

This is useful in deriving the Boltzmann machine algorithm. Free energy is like an exponentially weighted sum of energies. That is $\exp(-F_E/k_B T) = Z = \Sigma_v \exp(-\beta\phi_v)$ and $\exp(-F_E/k_B T)/Z = \Sigma_v \exp(-\phi_v T)/Z = \Sigma_v p_v = 1$ depicts the sum of the probability of the states (which is just 1).

A.18 Entropy

The difference between average energy $<\phi_v>$ and the free energy F_E is given by:

$$
\begin{aligned}
<\phi_v> - F_E &= \Sigma_v p_v \phi_v - k_B T \log(Z), \quad &(\Sigma_v p_v = 1) \\
&= k_B T \Sigma_v p_v[\beta\phi_v - \log(Z)] \\
&= -k_B T \Sigma_v p_v \log[\exp(-\beta\phi_v/Z)] \\
&= -k_B T \Sigma_v p_v \log(p_v) \tag{A.29}
\end{aligned}
$$

The above expression, except for the k_BT term depicts a quantity called *entropy* $\mathcal{H}$. That is:

$$\mathcal{H} = -\Sigma\, p_v \log(p_v) \tag{A.30a}$$

and in terms of $\mathcal{H}$, the free-energy can be written as

$$F_E = <\phi_v> - k_BT\mathcal{H} \tag{A.30b}$$

The entropy refers to:

- Width of the probability distribution p_v.
- Larger $\mathcal{H}$ corresponding to more states v that have appreciable probability.
- Average amount of additional information required to specify one of the states; or larger entropy, to the more uncertain of the actual state v.

APPENDIX B

Matrix Methods in Little's Model

B.1   A Short Review on Matrix Theory

B.1.A   Left and Right Eigenvectors

Assuming A is an $M \times M$ square matrix, it has eigenvectors. Denoting $v^{(\ell)}$ and $v^{(r)}$ as the left (row) and right (column) eigenvectors, then:

$$v^{(\ell)}A = \lambda^{(\ell)}v^{(\ell)} \tag{B.1}$$

and

$$Av^{(r)} = \lambda^{(\ell)}v^{(\ell)} \tag{B.2}$$

where λ is a scalar known as the eigenvalue.

It should be noted that $v^{(r)}$ and $v^{(\ell)}$ are *orthogonal* unless $\lambda^{(r)} = \lambda^{(\ell)}$.

From Equation (B.2) it follows that:

$$(A - \lambda I_M)v^{(r)} = 0 \tag{B.3}$$

where I_M is called the *identity matrix* of order M.

Equation (B.3) refers to a system of homogeneous linear equations for which a nontrivial solution exists only if the determinant is zero. That is:

$$|A - \lambda I_M| = 0 \tag{B.4}$$

The same result of Equation (B.4) is also obtained for the eigenvalues of Equation (B.1), and hence the left and right eigenvectors have the same set of eigenvalues.

Last, Equation (B.3) can be used to find the eigenvectors, if the values of λ are known.

B.1.B   Degenerate and Nondegenerate Matrices

A matrix which has no two eigenvalues equal and, therefore, has just as many distinct eigenvalues as its dimension is said to be *nondegenerate*.

A matrix is degenerate if more than one *eigenvector* has the same *eigenvalue*.

A nondegenerate matrix can always be *diagonalized*. However, a degenerate matrix may or may not be put into a diagonal form, but it can always be reduced to what is known as the *Jordan normal* or *Jordan canonical form*.

B.1.C Diagonalizing a Matrix by Similarity Transform
Consider $\lambda_i (i = 1, ..., M)$ as the eigenvalues of A. Let v_i be an eigenvector of A associated with λ_i, $i = 1, 2, ..., M$. Then:

$$A v_i = \lambda_i v_i \qquad (B.5)$$

Step 1: Find the eigenvalues of A. If the eigenvalues are distinct (that is, when the matrix A is nondegenerate), then proceed to Step 2.
Step 2: Find eigenvectors of A.
Step 3: Construct the following matrix:

$$Q = [v_1 \ v_2 \ ... \ v_M] \qquad (B.6)$$

Step 4: Calculate $A = Q^{-1}AQ$, which refers to the *diagonal matrix*.

Note that if the eigenvalues of A are distinct, a diagonal matrix A can always be found. If, on the other hand, the eigenvalues of A are not all distinct (repeated eigenvalues), it still is possible to find A by implementing the above procedure if a linearly independent eigenvector can be found to correspond to each eigenvalue. However, when there are not as many linearly independent eigenvectors as eigenvalues, diagonalization is impossible.

B.1.D Jordan Canonical Form
When A cannot be transformed into a diagonal form (in certain cases if it has repeated eignvalues), it can however be specified in Jordan canonical form expressed as:

$$J = \begin{bmatrix} J_1 & 0 & . & . & . & 0 \\ 0 & J_2 & . & . & . & 0 \\ & & . & & & \\ & & . & & & \\ & & . & & & \\ 0 & 0 & . & . & . & J_r \end{bmatrix} \qquad (B.7)$$

in which J_i is an $(m_i \times m_i)$ matrix of the form:

$$\begin{bmatrix} \lambda & 1 & 0 & . & . & . & 0 \\ 0 & \lambda & 1 & . & . & . & 0 \\ 0 & 0 & \lambda & . & . & . & 0 \\ & & & . & & & \\ & & & . & & & \end{bmatrix} \qquad (B.8)$$

That is, J_i has all diagonal entries λ, all entries immediately. above the diagonal are 1's, and all other entries are 0's. J_i is called a *Jordan block* of size m_i .

The procedure for transforming A into Jordan form is similar to the method discussed in Section B.1.C, and the *generalized* (or *principal*) *eigenvectors* can be considered as follows.

A vector **v** is said to be a generalized eigenvector of grade k of A associated with λ iff:

$$\begin{cases} (A - \lambda I_M)^k v = 0 \\ (A - \lambda I_M)^{k-1} v \neq 0 \end{cases} \qquad (B.9)$$

$$v_{k-1} = (A - \lambda I_M) v_k, \quad v_{k-2} = (A - \lambda I_M) v_{k-1}$$

$$Q = [v_1 \ v_2 \ ... \ v_M]$$

Claim: Given an $M \times M$ matrix A, let λ be an eigenvalue of *multiplicity* $\mathfrak{m}$, then one can always find $\mathfrak{m}$ linearly independent generalized eigenvectors associated with λ .

B.1.E Generalized Procedure to Find the Jordan Form
- Factorize: $\det(\lambda I_M - A) = (\lambda - \lambda_1)^{k_1} \ ... \ (\lambda - \lambda_r)^{k_r}$
$$\Sigma k_i = M$$
for each λ_i, $i = 1, 2, 3 ..., r$.

- Recall: A null space of a linear operator A is the set N(A) defined by:

$$N(A) = \{ \text{all the elements x of } (F^n, F) \text{ for which } Ax = 0 \}$$
$$v(A) = \text{rank of } N(A) = \text{nullity of } A$$

Theorem B.1.E.1: Let A be an $m \times n$ matrix. Then $\rho(A) + v(A) = n$, where $\rho(A)$ is the rank of A equal to the maximum number of linearly independent columns of A.

• Proceed as per the following steps:

Step 1: Find $\mathfrak{m}_i$ denoting the smallest $\mathfrak{m}$ such that $v(A - \lambda_i I_M)^{\mathfrak{m}} = k_i \bullet \mathfrak{m}_i$ is equal to the length of the longest chains, and v signifies the nullity.

Step 2: Find a basis for $N[(A - \lambda_i I_M)]^{\mathfrak{m}_i}$ that does not lie in $N[(A - \lambda_i I_M)]^{\mathfrak{m}_i-1}$.

Step 3: Find the next vector in each chain already started by $v_{j,\mathfrak{m}_i-1}$. These lie in $N[(A - \lambda_i I_M)]^{\mathfrak{m}_i-1}$ but not in $N[(A - \lambda_i I_M)]^{\mathfrak{m}_i-2}$.

Step 4: Complete the set of vectors as a basis for the solutions of $(A - \lambda_i I_M)^{\mathfrak{m}_i-1} v = 0$ which, however, do not satisfy $(A - \lambda_i I_M)^{\mathfrak{m}_i-2} v = 0$.

Step 5: Repeat Steps (3) and (4) for successively higher levels until a basis (of dimension k_i) is found for $N[(A - \lambda_i I_M)]^{\mathfrak{m}_i}$.

Step 6: Now the complete structure of A relative to λ_i is known.

B.1.F Vector Space and Inner Product Space

A set V whose operations satisfy the list of requirements specified below is said to be a *real vector space*. V is a set for which addition and scalar multiplication are defined and $x, y, z \in V$. Further, α and β are real numbers. The axioms of a vector space are:

1) $x + y = y + x$
2) $(x + y) + z = x + (y + z)$
3) There is an element in V, denoted by 0, such that $0 + x = x + 0 = x$.
4) For each x in V, there is an element $-x$ in V such that $x + (-x) = (-x) + x = 0$.
5) $(\alpha + \beta)x = \alpha x + \beta x$
6) $\alpha(x + y) = \alpha x + \alpha y$
7) $(\alpha\beta)x = \alpha(\beta x)$
8) $1.x = x$

An inner product of two vectors in R^M denoted by (a, b) is the real number $(a, b) = \alpha_1\beta_1 + \alpha_2\beta_2 + ... + \alpha_M\beta_M$. If V is a real vector space, a function that assigns to every pair of vectors x and y in V a real number (x, y) is said to be an inner product on V, if it has the following properties:

1. $(x, x) \geq 0$
$(x, x) = 0$ iff $x = 0$
2. $(x, y + z) = (x, y) + (x, z)$
$(x + y, z) = (x, z) + (y, z)$
3. $(\alpha x, y) = \alpha(x, y)$
$(x, \beta y) = \beta(x, y)$
4. $(x, y) = (y, x)$

B.1.G Symmetric Matrices

If A is an $M \times N$ matrix, then $A = [a_{ij}]_{(MN)}$ is *symmetric* if $A = AT$, where $AT = [b_{ij}]_{(NM)}$ and $b_{ij} = a_{ji}$

If U is a real $M \times M$ matrix and $U^T U = I_M$, U is said to be an *orthogonal matrix*. Clearly, any orthogonal matrix is invertible and $U^{-1} = U^T$, where U^T denotes the *transpose* of U.

Two real matrices C and D are said to be *similar* if there is a real invertible matrix B such that $B^{-1}CB = D$. If A and B are two $M \times M$ matrices and there is an orthogonal matrix U such that $A = U^{-1}BU$, then A and B are orthogonally similar. A real matrix is diagonalizable over the "reals", if it is similar over the reals to some real diagonal matrix. In other words, an $M \times M$ matrix A is diagonalizable over the reals, iff R^M has a *basis** consisting of eigenvectors of A.

Theorem B.1.G.1: Let A be an $M \times M$ real symmetric matrix. Then any eigenvalue of A is real.

Theorem B.1.G.2: Let T be a symmetric linear operator (that is, $T(x) = Ax$) on a finite dimensional inner product space V. Then V has an orthonormal basis of eigenvectors of T.

As a consequence of Theorem B.1.G.2, the following is a corollary: If A is a real symmetric matrix, A is orthogonally similar to a real diagonal matrix.

It follows that any $M \times M$ symmetric matrix is diagonalizable.

---

* Definition: Let V be a vector space and $\{x_1, x_2, ..., x_M\}$ be a collection of vectors in V. The set $\{x_1, x_2, ..., x_M\}$ is said to be a *basis* for V if, (1) $\{x_1, x_2, ..., x_M\}$ is a linearly independent set of vectors; and (2) $\{x_1, x_2, ..., x_M\}$ spans V.

B.2   Persistent States and Occurrence of Degeneracy in the Maximum Eigenvalue of the Transition Matrix

As discussed in Chapter 5, Little in 1974 [33] defined a $2^M \times 2^M$ matrix T_M whose elements give the probability of a particular state $|S_1, S_2, ..., S_M>$ yielding after one cycle the new state $|S'_1, S'_2, ..., S'_M>$.

Letting $\Psi(\alpha)$ represent the state $|S_1, S_2, ..., S_M>$ and $\Psi(\alpha')$ represent $|S'_1, S'_2, ..., S'_M>$, the probability of obtaining a configuration after m cycles is given by:

$$\Psi(\alpha') \, T_M^m \, \Psi(\alpha) \tag{B.10}$$

If T_M is diagonalizable, then a representation of $\Psi(\alpha)$ can be made in terms of eigenvectors alone. These eigenvectors would be linearly independent. Therefore, as referred to in Section B.1, assuming T_M is diagonalizable, from Equation (B.5) it can be stated that:

$$T_M \Psi(\alpha) = \lambda_r \vartheta_r(\alpha) \tag{B.11}$$

where ϑ_r are the eigenvectors of the operator T_M. There are 2^M such eigenvectors each of which has 2^M components $\vartheta_r(\alpha)$, one for each configuration α; λ_r is the rth eigenvalue.

However, as discussed in Section B.1, if T_M were symmetric, the assumption of its diagonalizability is valid since any symmetric matrix is diagonalizable. However, as stated previously, it is reasonably certain that T_M may not be symmetric. Initially Little assumes that T_M is symmetric and then shows how his argument can be extended to the situation in which T_M is not diagonalizable.

Assuming T_M is diagonalizable, $\Psi(\alpha)$ can be represented as:

$$\Psi(\alpha) = \sum_{r=1}^{2^M} \vartheta_r(\alpha) \tag{B.12}$$

as is familiar in quantum mechanics; and assuming that the $\vartheta_r(\alpha)$ are normalized to unity, it follows that:

$$\sum_{\alpha} \vartheta_r(\alpha) \vartheta_r(\alpha') = \delta_{rs} \tag{B.13}$$

where $\delta_{rs} = 0$, if $r \neq s$ and $\delta_{rs} = 1$, if $r = s$; and noting that $\Psi(\alpha')$ can be expressed as $\sum_{r=1}^{2^M} \vartheta_r(\alpha')$ similar to Equation (B.13) so that:

$$\langle \Psi(\alpha')|T_M|\Psi(\alpha)\rangle = \sum_r \lambda_r \vartheta_r(\alpha')\vartheta_r(\alpha) \tag{B.14}$$

Now the probability $\Gamma(\alpha_1)$ of obtaining the configuration α_1 after m cycles can be elucidated. It is assumed that after M_o cycles, the system returns to the initial configuration and averaging over all initial configurations:

$$\Gamma(\alpha_1) = \sum_\alpha \langle a\, |T_M^{M_o-m}|\, a_1\rangle \langle a_1\, |T_M^{M_o-m}|\, a\rangle / \sum_\alpha \langle\alpha|T_M|\alpha\rangle \tag{B.15}$$

Using Equation (B.14), Equation (B.15) can be simplified to:

$$\Gamma(\alpha_1) = \sum_\alpha \sum_{ru} \vartheta_u(\alpha)\lambda_u^{M_o-m}\vartheta_u(\alpha_1)\vartheta_r(\alpha_1)\lambda_r^m\vartheta_r(\alpha)/\sum_r \lambda_r^{M_o} \tag{B.16}$$

Then, by using Equation (B.13), Equation (B.15) can further be reduced to:

$$\Gamma(\alpha_1) = \sum_r \lambda_r^{M_o}\vartheta_r^2(\alpha)/\sum_r \lambda_r^{M_o} \tag{B.17}$$

Assuming the eigenvalues are nondegenerate and M_o is a very large number, then the only significant contribution to Equation (B.17) comes from the maximum eigenvalues. That is:

$$\sum_r \lambda_r^{M_o} \approx \lambda_{max}^{M_o} \tag{B.18}$$

$$\Gamma(\alpha_1) \approx \lambda_{max}^{M_o}\,\vartheta_{max}^2(\alpha_1)/\lambda_{max}^{M_o}$$

$$= \vartheta_{max}^2(\alpha_1) \tag{B.19}$$

However, if the maximum eigenvalue of T_M is degenerate or sufficiently close in value (for example $|\lambda_1^{M_o}| \approx |\lambda_2^{M_o}|$), then these degenerate eigenvalues contribute and Equation (B.17) becomes:

$$\Gamma(\alpha_1) = [\lambda_1^{M_o}\vartheta_1^2(\alpha_1) + \lambda_2^{M_o}\vartheta_2^2(\alpha_1)\,]/(\lambda_1^{M_o} + \lambda_2^{M_o}) \tag{B.20}$$

216

In an almost identical procedure to that as discussed above, Little found the probability of obtaining a configuration α_2 after ℓ cycles as:

$$\Gamma(\alpha_1, \alpha_2) = \sum_{ru} \lambda_r^{M_o-\ell+m} \lambda_u^{\ell-m} \vartheta_r(\alpha_2)\vartheta_u(\alpha_2)\vartheta_u(\alpha_1)\vartheta_r(\alpha_1) / \sum_r \lambda_r^{M_o} \tag{B.21}$$

$$= \vartheta_{max}^2(\alpha_1)\vartheta_{max}^2(\alpha_2) \tag{B.22}$$

if the eigenvalues are degenerate and M_o, $(\ell - m)$ are large numbers. Examining Equation (B.19), one finds:

$$\Gamma(\alpha_1, \alpha_2) = \Gamma(\alpha_1)\Gamma(\alpha_2) \tag{B.23}$$

Thus, the influence of configuration α_1 does not affect the probability of obtaining the configuration α_2.

However, considering the case where $|\lambda_1^{M_o}| \approx |\lambda_2^{M_o}|$.

$$\begin{aligned}
\Gamma(\alpha_1, \alpha_2) = &\{\lambda_1^{M_o}\vartheta_1^2(\alpha_2)\vartheta_1^2(\alpha_1) + \lambda_2^{M_o}\vartheta_2^2(\alpha_2)\vartheta_2^2(\alpha_1) \\
&+ \lambda_1^{M_o-\ell+m}\lambda_2^{\ell-m}\vartheta_1(\alpha_2)\vartheta_2(\alpha_2)\vartheta_2(\alpha_1)\vartheta_1(\alpha_1) \\
&+ \lambda_2^{M_o-\ell+m}\lambda_1^{\ell-m}\vartheta_2(\alpha_2)\vartheta_1(\alpha_2)\vartheta_1(\alpha_1)\vartheta_2(\alpha_1)\} \bullet \\
&\{\lambda_1^{M_o} + \lambda_2^{M_o}\}^{-1}
\end{aligned} \tag{B.24}$$

The probability of obtaining a configuration α_2 is then dependent upon the configuration α_1, and thus the influence of α_1 persists for an arbitrarily long time.

B.3 Diagonalizability of the Characteristic Matrix

It was indicated earlier that Little assumed T_M is diagonalizable. However, his results also can be generalized to any arbitrary $M \times M$ matrix because any $M \times M$ matrix can be put into Jordan form as discussed in Section B.1.D.

In such a case, the representation of $\Psi(\alpha)$ is made in terms of generalized eigenvectors which satisfy:

$$(T_M - \lambda_r I_M)^k t_r(\alpha) = 0 \tag{B.25}$$

instead of pertaining to the eigenvectors alone as discussed in Section B.1.D.

The eigenvectors are the generalized vectors of grade 1. For $k > 1$, Equation (B.25) defines the generalized vectors. Thus for a particular case of $k = 1$, the results of Equation (B.14) and Equation (B.16) are the same. Further, an eigenvector $\Psi(\alpha)$ can be derived from Equation (B.25) as follows:

$$\Psi(\alpha) = (T_M - \lambda_r I_M)^{k-1} t_r(\alpha)/(k-1)! \tag{B.26}$$

Lemma. Let B be an $M \times M$ matrix. Let p be a principal vector of grade $g \geq 1$ belonging to the eigenvalue μ. Then for large k one has the asymptotic forms

$$B^k p = k^{g-1}\mu^k \mathfrak{v} + r^{(k)} \tag{B.27}$$

where $\mathfrak{v}$ is an eigenvector belonging to μ and where the remainder $r^{(k)} = 0$ if $g = 1$ or:

$$r^{(k)} \Rightarrow ⊕ (k^{g-2}|\mu|^k), \quad \text{if } g > 2; \quad (⊕ \Rightarrow \text{order of}) \tag{B.28}$$

For the general case one must use the asymptotic form for large m. Hence, in the present case:

$$T_M^m t_r(\alpha) = m^{k-1} \lambda^m \Psi(\alpha) + r^{(m)} \tag{B.29}$$

where $r^{(m)} \approx m^{k-2}|\lambda|^m$, which yields:

$$\langle \alpha'|T_M^m|\alpha\rangle \approx \sum_{rk} m^{k-1}\lambda_r^m \Psi_{rk}(\alpha')\Psi_{rk}(\alpha) \tag{B.30}$$

where $\Psi_{rk}(\alpha)$ is the eigenvector of eigenvalue λ_r for the generalized vector $t_r(\alpha)$ defined in Equation (B.26). By using this form in Equation (B.15), the following is obtained:

$$\Gamma(\alpha_1) = \sum_{rk} m^{2k-2}\lambda_r^m \Psi_{rk}^2(\alpha_1)/\sum_r \lambda_r^{M_0} \tag{B.31}$$

The above relation is dependent on m if there is a degeneracy in the maximum eigenvalues. It means that there is an influence from constraints or knowledge of the system at an earlier time. Also, from Equation (B.13) it is required that the generalized vectors must be of the same grade. That is, from Equation (B.15) it follows that:

$$\Gamma(\alpha_1) = \sum_{\alpha} \sum_{rku} m^{k-1}\Psi_u(\alpha)\lambda_u^{M_0-m}\Psi_u(\alpha_1)m^{k-1}\Psi_{rk}(\alpha_1)\lambda_r^m\Psi_{rk}(\alpha)/\sum_r \lambda_r^{M_0}$$

$\neq 0$ iff $r, k = u$ $\hspace{5cm}$ (B.32)

Therefore, the conditions for the persistent order are: The maximum eigenvalues must be degenerate, and their generalized vectors must be of the same grade.

APPENDIX C

Overlap of Replicas and Replica Symmetry Ansatz

In a collective computational task, the simplest manner by which computation is accomplished refers to the associative memory problem stated as follows.

When a set of $\mathcal{P}$ patterns $\{\xi_i^\mu\}$, labeled by ($\mu = 1, 2, ..., \mathcal{P}$), is stored in a network with N interconnected units (designated by $i = 1, 2, ..., N$), the network responds to deliver whichever one of the stored patterns most closely resembles a new pattern ξ_i presented at the network's input.

The network stores a stable pattern (or a set of patterns) through the adjustment of its connection weights. That is, a set of patterns $\{\xi_i^\mu\}$ is presented to the network during a training session and the connection strengths (W_{ij}) are adjusted on a correlatory basis *via* a superposition of terms:

$$W_{ij} = (1/N) \sum_{\mu=1}^{\mathcal{P}} \xi_i^\mu \xi_j^\mu \tag{C.1}$$

Such an adjustment calls for a minimization of energy functional when the *overlap* between the network configuration and the stored pattern ξ_i is largest. This energy functional is given by the Hamiltonian:

$$\mathcal{H}_N = -(1/2N) \sum_{\mu=1}^{\mathcal{P}} \sum_i (S_i \xi_j^\mu)^2$$

$$\Rightarrow -(1/2) \sum_{ij} [(1/N) \sum_{\mu=1}^{\mathcal{P}} (\xi_i^\mu \xi_j^\mu)] S_i S_j \tag{C.2}$$

In the event minimization is not attained, the residual overlap between ξ_i^μ and the other patterns gives rise to a *cross-talk* term. The cross-talk between different patterns on account of their random overlap would affect the recall or retrieval of a given pattern, especially when $\mathcal{P}$ becomes of the order of N.

To quantify the random overlaps, one can consider the average free energy associated with the random binary pattern. Implicitly it refers to averaging the log(Z), but computation of <<log(Z)>> is not trivial. Therefore, log(Z) is specified by the relation:

$$\log(Z) = \lim_{n \to 0} [(Z^n - 1)/n] \tag{C.3}$$

and the corresponding averaging would involve Z^n and not $\log(Z)$. The quantity Z^n can be considered as the partition function of n *copies* or *replicas* of the original system. In other words:

$$Z^n = T_{S1}^{\Sigma} T_{S2}^{\Sigma} \dots T_{Sn}^{\Sigma} \exp[-\beta(E\{S_i^1\} + \dots + E\{S_i^n\})] \tag{C.4}$$

where each replica is indicated by a superscript (*replica index*) on its S_i's running from 1 to n.

In the conventional method of calculating $<<Z^n>>$ *via* saddle-point technique, the relevant order parameters (overlap parameters) derived at the saddle points can be assumed to be symmetric in respect to replica indices. That is, the saddle-point values of the order parameters do not depend on their replica indices. This is known as *replica symmetry ansatz* (hypothesis).

This replica symmetry method, however, works only in certain cases where reversal of limits is justified and $<<Z^n>>$ is calculated for *integer n* (eventually interpreted n as a real number). When it fails, a more rigorous method is pursued with *replica symmetry breaking ansatz*.

BIBLIOGRAPHY

[1]   Müller, B., and Reinhardt, J.: *Neural Networks: An Introduction* (Springer-Verlag, Berlin: 1990).

[2]   Gerstein, G. L., and Mandelbrot, B.: "Random walk models for the spike activity of a single neuron." *Biophys. J.*, 4, 1964, 41-68.

[3]   Feinberg, S. E., and Hochman, H. G.: "Modal analysis of renewal models for spontaneous single neuron discharges." *Biol. Cybern.*, 11, 1972, 201-207.

[4]   Neelakanta, P. S., Mobin, M. S., Pilgreen, K., and Aldes, L.: "Markovian dichotomous model of motor unit action potential trains: EMG analysis." Proc. 9th Annual Pittsburgh Conf. (Modelling and Simulation), (May 5-6, 1988, Pittsburgh, PA), 19, Part 4, 1735-1739.

[5]   Neelakanta, P. S., Mobin, M. S., Pilgreen, K., and Aldes, L.: "Resolution of power spectral analysis of events in the electromyogram: An error estimation model." Proc. 9th Annual Pittsburgh Conf. (Modelling and Simulation), (May 5-6, 1988, Pittsburgh, PA), 19, Part 5, 2421-2425.

[6]   Sampath, G., and Srinivasan, S. K.: *Stochastical Models for Spike Trains of Single Neurons* (Springer-Verlag, Berlin: 1977).

[7]   McCulloch, W. W., and Pitts, W.: "A logical calculus of the ideas imminent in nervous activity." *Bull. Math. Biophys.*, 5, 1943, 115-133.

[8]   MacGregor, R. J.: *Neural and Brain Modeling* (Academic Press, San Diego, CA: 1987), 13-33, 138-139, 140-143.

[9]   Wiener, N.: *Cybernetics: Control and Communication in the Animal and the Machine* (MIT Press, Cambridge, MA: 1948 & 1961).

[10]   Gabor, D.: "Theory of communication." *J. Inst. Electr. Eng.*, 93, 1946, 429-457.

[11]   Griffith, J. S.: "A field theory of neural nets I." *Bull. Math. Biophys.*, 25, 1963, 11-120.

[12]   Griffith, J. S.: "A field theory of neural nets II." *Bull. Math. Biophys.*, 27, 1965, 187-195.

[13]   Griffith, J. S.: "Modelling neural networks by aggregates of interacting spins." *Proc. R. Soc. London*, A295, 1966, 350-354.

[14]   Griffith, J. S.: *Mathematical Neurobiology* (Academic Press, New York: 1971).

[15]   De Groff, D., Neelakanta, P. S., Sudhakar, R., and Medina, F.: "Liquid-crystal model of neural networks." *Complex Syst.*, 7, 1993, 43-57.

[16]   Neelakanta, P. S., Sudhakar, R., and De Groff, D.: "Langevin machine: A neural network based on stochastically justifiable sigmoidal function." *Bio. Cybern.*, 65, 1991, 331-338.

[17]   De Groff, D., Neelakanta, P. S., Sudhakar, R., and Medina, F.: "Collective properties of neuronal activity: Momentum flow and particle dynamics representation." *Cybernetica*, XXXVI, 1993, 105-119.

[18]   De Groff, D., Neelakanta, P. S., Sudhakar, R., and Aalo, V.: "Stochastical aspects of neuronal dynamics: Fokker-Planck approach." *Bio. Cybern.*, 69, 1993, 155-164.

[19]   Hebb, D. O.: *The Organization of Behavior* (Wiley and Sons, New York: 1949).

[20]   Stanley-Jones, D., and Stanley-Jones, K.: *Kybernetics of Natural Systems, A Study in Patterns of Control* (Pergamon Press, London: 1960).

[21]   Uttley, A. M.: "The classification of signals in the nervous system." *Electroencephalogr. Clin. Neurophysiol.*, 6, 1954, 479.

[22]   Shannon, C. E.: "Synthesis of two-terminal switching circuits." *Bell Syst. Tech. J.*, 28, 1949, 656-715.

[23]   Walter, W. G.: *The Living Brain* (Duckworth, London: 1953).

[24]   Ashby, W. R.: *An Introduction to Cybernetics* (Chapman and Hall, London: 1956).

223

[25]   George, F. H.: *Cybernetics and Biology* (W. H. Freeman and Co., San Francisco: 1965).

[26]   Arbib, M. A.: *Brains, Machines and Mathematics* (McGraw-Hill Book Co., New York: 1964).

[27]   Kohonen, T.: *Self-organization and Associative Memory* (Springer-Verlag, Berlin: 1989).

[28]   Hodgkin, A. L., and Huxley, A. F.: "A quantitative description of membrane current and its application to conduction and excitation in nerve." *J. Physiol. (London)*, 117, 1952, 500-544.

[29]   Agnati, L. F., Bjelke, B., and Fuxe, K.: "Volume transmission in the brain." *Am. Scientist*, 80, 1992, 362-373.

[30]   Turing, A. M.: "On computable numbers, with an application to the Entscheidungs problem." *Proc. London. Math. Soc. Ser. 2*, 42, 1937, 230-265.

[31]   Hopfield, J. J.: "Neural networks and physical systems with emergent collective computational abilities." *Proc. Natl. Acad. Sci. U.S.A.*, 79, 1982, 2554-2558.

[32]   Cragg, B. G., and Temperley, H. N. V.: "The organization of neurons: A cooperative analogy." *Electroencephalogr. Clin. Neurophysiol.*, 6, 1954, 85-92.

[33]   Little, W. A.: "The existence of persistent states in the brain." *Math. Biosci.*, 19, 1974, 101-120.

[34]   Hopfield, J.J., and Tank, D.W.: " 'Neural' computation of decision in optimization problems." *Biol. Cybern.*, 52, 1985, 1-12.

[35]   Ising, E.: Ph.D. dissertation, *Z. Phys.*, 31, 1925, 253.

[36]   Hopfield, J. J.: "Neurons with graded response have collective properties like those of two-state neurons." *Proc. Natl. Acad. Sci. U.S.A.*, 81, 1984, 3088-3092.

[37]   Thompson, R. S., and Gibson, W. G.: "Neural model with probabilistic firing behavior. I. General considerations." *Math. Biosci.*, 56, 1981a, 239-253.

[38]   Peretto, P. "Collective properties of neural networks: A statistical physics approach." *Biol. Cybern.*, 50, 1984, 51-62.

[39]   Rosenblatt, F.: "The perceptron: A probabilistic model for information storage and organization in the brain," *Psychol. Rev.*, 65, 1958, 42 & 99.

[40]   Rosenblatt, F.: *Principles of Neurodynamics: Perceptrons and the Theory of Brain Mechanisms* (Spartan Books, Washington., D. C.: 1961).

[41]   Anderson, J. A.: "Cognitive and psychological computation with neural model." *IEEE Trans. Syst. Man Cybern.*, SCM-13, 1983, 799-815.

[42]   Beurle, R. L.: "Properties of a mass of cells capable of regenerating pulses." *Philos. Trans. R. Soc. London Ser. A*, 240, 1956, 55-94.

[43]   Wilson, H.R., and Cowan, J.D.: A mathematical theory of the functional dynamics of cortical thalmic nervous tissue." *Biol. Cybern.*, 3, 1973, 55-80.

[44]   Wilson, H. R., and Cowan, J. D.: "Excitatory and inhibitory interactions in localized populations of model neurons." *Biophys. J.*, 12, 1972, 1-24.

[45]   Oguztoreli, M. N.: "On the activities in a continuous neural network." *Biol. Cybern.*, 18, 1975, 41-48.

[46]   Kawahara, T., Katayama, K., and Nogawa, T.: "Nonlinear equations of reaction-diffusion type for neural populations." *Biol. Cybern.*, 48, 1983, 19-25.

[47]   Ventriglia, F.: "Propagation of excitation in a model of neuronal systems." *Biol. Cybern.*, 1978, 30, 75-79.

[48]   Kung, S. Y.: *Digital Neural Networks* (Prentice Hall, Englewood Cliffs, NJ: 1993).

[49]   Bergstrom, R. M., and Nevalinna, O.: "An entropy model of primitive neural systems." *Int. J. Neurosci.*, 4, 1972, 171-173.

[50]   Takatsuji, M.: "An information-theoretical approach to a system of interacting elements." *Biol. Cybern.*, 17, 1975, 207-210.

[51]   Hinton, G. E., Sejnowski, T. J., and Ackley, D. H.: "Boltzmann machines: Constraint satisfaction networks that learn." Tech. Rep. SMU-CS-84-119 (Carnegie-Mellon University, Pittsburgh, PA: 1984).

[52]   Aarts, E., and Korst, J.: *Simulated Annealing and Boltzmann Machines* (John Wiley and Sons, Chichester: 1989).

[53]   Szu, H., and Hartley, R.: "Fast simulated annealing." *Phys. Lett. A.*, 122, 1987, 157-162.

[54]   Akiyama, Y., Yamashita, A., Kajiura, M., and Aiso, H.: "Combinatorial optimization with Gaussian machines." Proc. Int. Joint Conf. Neural Networks (June 18-22, 1990, Washington, D. C.), I 533-I 539.

[55]   Geman, S, and Geman, D.: "Stochastic relaxation, Gibbs distributions and Bayesian restoration of images," *IEEE Trans. Pattern Anal. Mach. Intell.*, 6, 1984, 721-741.

[56]   Jeong, H., and Park, J. H.: "Lower bounds of annealing schedule for Boltzmann and Cauchy machines." Proc. Int. Joint Conf. Neural Networks (June 18-22, 1990, Washington, D. C.), I 581-I 586.

[57]   Ackley, D. H., Hinton, G. E., and Sejnowski, T. J.: "A learning algorithm for Boltzmann machines." *Cognit. Sci.*, 9, 1985, 147.

[58]   Liou, C. Y., and Lin, S. L.: "The other variant Boltzmann machine." Proc. Joint Conf. Neural Networks (June 18-22, 1990, Washington, D. C.), I 449- 454.

[59]   Feldman, J. A., and Ballard, D. H.: "Connectionist models and their properties." *Cognit. Sci.*, 6, 1982, 205-254.

[60]   Livesey, M.: "Clamping in Boltzmann machines." *IEEE Trans. Neural Networks*, 2, 1991, 143-148.

[61]   Györgi, G., and Tishby, N.: "Statistical Theory of learning a rule." *Proc. Stat. Phys 17 Workshop on Neural Networks and Spin Glasses* (August 8-11, 1989, Porto Alegre, Brazil). (Eds.

Theumann, W. K., and R. Köberle) (World Scientific, Singapore: 1990).

[62]   Personnaz, L.,Guyon, I., and Dreyfus, G.: "Collective computational properties of neural networks: New learning mechanisms." *Phys. Rev. A*, 34, 1989, 4303.

[63]   Kanter, I., and Sompolinsky, H: "Associative recall of memories without errors." *Phys. Rev. A*, 35, 1987, 380.

[64]   Caianello, E. R.: "Outline of thought processes and thinking machines." *J. Theor. Biol.*, 2, 1961, 204.

[65]   Cowan, J. D.: "Statistical mechanics of nervous nets." In *Neural Networks* (Ed. Caianello, E.R.), (Springer-Verlag, Berlin: 1968).

[66]   Amari, S.: "A method of statistical neurodynamics." *Biol. Cybern.*, 14, 1974, 201-215.

[67]   Ingber, L.: "Statistical mechanics of neocortical interactions. I. Basic formulation." *Phys. Rev. D*, 5D, 1982, 83-107.

[68]   Ingber, L.:"Statistical mechanics of neocortical interactions. Dynamics of synaptic modifications." *Phys. Rev. A*, 8, 1983, 395-416.

[69]   Amit, D. J., Gutfreund, H., and Sompolinsky, H: "Spin-glass models of neural networks." *Phys. Rev.*, 32, 1985a, 1007-1018.

[70]   Toulouse, G., Dehaene, S., and Changeux, J. P.: "Spin glass model of learning by selection." *Proc. Natl. Acad. Sciences U.S.A.*, 83, 1986, 1695-1698.

[71]   Rumelhart, D. E., Hinton, G. E., and Williams, R. J.: "Learning representations by back-propagating errors." *Nature* (London), 323, 1986, 533-536.

[72]   Gardner, E.: "Maximum storage capacity in neural networks." *Europhys. Lett.*, 4, 1987, 481-485.

[73]   Theumann, W. K., and Koberle, R. (Eds.): "Neural networks and spin glasses." *Proc. Stat. Phys. 17 Workshop on Neural Networks and Spin Glasses* (August 8-11, 1989, Porto Alegre, Brazil), (World Scientific, Singapore: 1990).

227

[74] MacGregor, R. J., and Lewis, E. R.: *Neural Modeling* (Plenum Press, New York: 1977).

[75] Licklider, J. C. R.: "Basic correlates of auditory stimulus." In *Stevens Handbook of Experimental Psychology* (S. S. Stevens, Ed., John Wiley, New York: 1951), 985-1039.

[76] Shaw, G. L., and Vasudevan, R.: "Persistent states of neural networks and the random nature of synaptic transmission." *Math. Biosci.*, 21, 1974, 207-218.

[77] Little, W. A., and Shaw, G. L.: "Analytic study of the memory storage capacity of a neural network." *Math. Biosci.*, 39, 1978, 281-290.

[78] Little, W. A., and Shaw, G. L.: "A statistical theory of short and long term memory." *Behav. Biol.*, 14, 1975, 115-133.

[79] Thompson, R. S., and Gibson, W. G.: "Neural model with probabilistic firing behavior. II. One-and two neuron networks." *Math. Biosci.*, 56, 1981b, 255-285.

[80] Toulouse, G.: "Symmetry and topology concepts for spin glasses and other glasses." *Phys. Rep.*, 49, 1979, 267.

[81] Brown, G. H., and Wolken, J. J.: *Liquid Crystals and Biological Structures* (Academic Press, New York: 1979).

[82] Chandrasekhar, S.: *Liquid Crystals* (University Press, Cambridge: 1977).

[83] Wannier, G.H.: *Statistical Physics* (Dover Publications, New York: 1966).

[84] Stornetta, W. S., and Huberman, B. A.: "An improved three-layer back propagation algorithm." *Proc. of the IEEE First Int. Conf. Neural Networks* (Eds. M. Caudill, and C. Butler, SOS Printing, San Diego, CA: 1987).

[85] Indira, R., Valsakumar, M. C., Murthy, K. P. N., and Ananthakrishnan, G.: "Diffusion in bistable potential: A comparative study of different methods of solution." *J. Stat. Phys.*, 33, 1983, 181-194.

[86]   Risken, R.: *The Fokker-Planck Equation* (Springer, Berlin: 1984).

[87]   Chandrasekhar, S.: "Stochastic problems in physics and astronomy." *Rev. Mod. Phys.*, 15, 1943, 1-89.

[88]   Papoulis, A.: *Probability, Random Variables and Stochastic Processes* (McGraw-Hill, New York: 1984) 392.

[89]   Valsakumar, M. C.: "Unstable state dynamics: Treatment of colored noise." *J. Stat. Phys.*, 39, 1985, 347-365.

[90]   Bulsara, A. R., Boss, R. D., and Jacobs, E. W.: "Noise effects in an electronic model of a single neuron." *Biol. Cybern.*, 61, 1989, 211-222.

[91]   Yang, X., and Shihab, A. S.: "Minimum mean square error estimation of connectivity in biological neural networks." *Biol. Cybern.*, 65, 1991, 171-179.

[92]   Yuan, J., Bhargava, A. K., and Wang, Q.: "An error correcting neural network." Conf. Proc. IEEE Pacific Rim Conf. on Commn., Computers & Signal Processing (June 1-2, 1989, Victoria, BC, Canada) 530-533.

[93]   Okuda, M., Yoshida, A., and Takahashi, K.: "A Dynamical behaviour of active regions in randomly connected neural networks." *J. Theor. Biol.*, 48, 1974, 51-73.

[94]   Pear, M. R., and Weiner, J. H.: "A Generalization of Kramer's rate formula to include some anharmonic effects." *J. Chem. Phys.*, 98, 1978, 785-793.

[95]   Marinescu, N., and Nistor, R.: "Quantum features of microwave propagation in rectangular waveguide." *Z. Naturforsch.*, 45a, 1990, 953-957.

[96]   Dirac, P.: *The Principles of Quantum Mechanics* (Clarendon Press, Oxford: 1987).

[97]   Amari, S.: "On mathematical models in the theory of neural networks." *Proc. First Int. Conf. Neural Networks*, 3, 1987, 3-10.

[98]   Abu-Mostafa, Y.S., and St. Jacques, S.: "Information capacity of the Hopfield model." *IEEE Trans. Inf., Theor.*, IT-31, 1985, 461-464.

[99]   McEliece, R. J., Posner, E. C., Rodemich, E. R., and Venkatesh, S. S.: "The capacity of the Hopfield associative memory." *IEEE Trans. Inf. Theor.*, IT-33, 1987, 461-482.

[100]   Weisbuch, G.: "Scaling laws for the attractors of the Hopfield networks." *J. Phys. Lett.*, 46, 1985, L623-L630.

[101]   Lee, K., Kothari, S. C., and Shin, D.: "Probabilistic information capacity of Hopfield associative memory." *Complex Syst.*, 6, 1992, 31-46.

[102]   Gardner, E., and Derrida, B.: "Optimal storage properties of neural network models." *J. Phys. A*, 21, 1988, 257-270.

[103]   Gardner, E., and Derrida, B.: "The unfinished works on the optimal storage capacity of networks." *J. Phys. A*, 22, 1989, 1983-1981.

[104]   Shannon, C. E., and Weaver, W.: *The Mathematical Theory of Communication* (University of Illinois Press, Urbana: 1949).

[105]   Bergstrom, R. M.: "An entropy model of the developing brain." *Dev. Psychobiol.*, 2, 1969, 139-152.

[106]   Uttley, A. M.: *Information Transmission in the Nervous System* (Academic Press, London: 1979).

[107]   Pfaffelhuber, E.: "Learning and information theory." *Int. Rev. Neurosci.*, 3, 1972, 83-88.

[108]   Legendy, C. R.: "On the scheme by which the human brain stores information." *Math. Biosci.*, 1, 1967, 555-597.

[109]   MacGregor, R. J. : *Theoretical Mechanics of Biological Neural Networks* (Academic Press Inc./Harcourt Brace Jovanovich Publishers, Boston : 1993).

[110]   MacGregor, R. J., and Gerstein, G. L.: "Cross-talk theory of memory capacity in neural networks." *Bio. Cybern.*, 65, 1991, 351-155.

[111]   Venikov, V. A. (Ed.): *Cybernetics in Electric Power Systems* (Mir Publishers, Moscow: 1978).

[112]   Aczel, J., and Daroczy, Z.: *On Measures of Information and Their Characteristics* (Academic Press, New York: 1975).

[113]   Kullback, S.: *Information Theory and Statistics* (Dover Publications, New York: 1968).

[114]   Lin, J.: "Divergence measures based on the Shannon entropy." *IEEE Trans. Inf. Theor.,* 37, 1991, 145-151.

[115]   Pospelov, D. A.: "Semiotic models in control systems." in *Cybernetics Today* (Ed. Makarov, I. M.), (Mir Publishers, Moscow: 1984).

[116]   Chowdhuri, D.: *Spin Glasses and Other Frustrated Systems* (Princeton University Press, Princeton, (NJ): 1986).

General Reading

[A]   Eccles, J. C.: *The Brain and the Unity of Conscious Experience* (Cambridge University Press, London: 1965).

[B]   Fuchs, W. R.: *Cybernetics for the Modern Mind* (The Macmillan Co., New York: 1971).

[C]   George, F. H.: *Cybernetics and Biology* (W. H. Freeman and Co., San Francisco: 1965).

[D]   Mammone, R. J., and Zeevi, Y. Y.: *Neural Networks : Theory and Applications* (Academic Press Inc., San Diego, CA.: 1991).

[E]   Nicolis, G., and Prigogine, I.: *Self-Organization in Nonequilibrium Systems* (John Wiley & Sons, New York: 1977).

[F]   Norwich, K. H.: *Infomation, Sensation and Perception* (Academic Press Inc., San Diego, CA.: 1993).

[G]   Pekelis, V.: *Cybernetic Medley* (Mir Publishers, Moscow: 1986).

[H]   Rose, J. (Ed.): *Progress in Cybernetics* (Vol. I&II) (Gordon and Breach Science Publishers, London: 1970).

231

Subject Index

T - #0126 - 101024 - C0 - 234/156/14 [16] - CB - 9780849324888 - Gloss Lamination